The Anthropology of Dragons: A Global Perspective

Jean S. Forward, Rachel M. Keeffe, Virginia McLaurin

Printed in the United States of America

ISBN 978-1-7329858-6-5

ACKNOWLEDGEMENTS

Jean:

First, A.J.Lent needs to be acknowledged. It was A.J. who sat in my office at the University of Massachusettes Amherst, one day, and convinced me that I should teach a one credit Honors course on Legends of the Beast (dragons). Several years later, an Honors student, Rachel M. Keeffe (now a graduate student in Biology and co-author of this textbook) convinced me to expand the course into a full 3 credit course that investigates the cultural, historical and ecological contexts of the over 400 dragon myths that she and Virginia "Jena" McLaurin (also a co-author) found in their research. The rest, as they say, is history.

Always I thank my family for their continual and consistent support of my love for dragons.

Most especially, my son, Tom, who was born in the year of the dragon.

Virginia:

I give great thanks to Jean and Rachel for inviting me in to work on this text, as well as the Department of Anthropology at UMass Amherst for their enthusiasm in supporting this project. I'd also like to thank my family for their support - my parents Erik and Michelle for giving me a love of stories, my husband Jon for talking through the research with me, and my kids Iris and Clell for asking me to tell these stories on camping trips, on car rides, and walking through supermarkets. It has been a pleasure to create this book - I will treasure our many conversations about dragons and look forward to many more!

Rachel:

Many thanks to Jean Forward for her excitement and support throughout this endeavor, as well as the Commonwealth Honors College at UMass Amherst for supporting my decision to make this project part of my honors thesis. I'd also like to acknowledge Alan Richmond for advising me on the diversity of reptiles and amphibians, which allowed me to think more deeply about the relationship between dragon myths and nature. And of course I would like to thank my family for always encouraging my love of art and science and the intersection of both.

CONTENTS

PREFACE

In creating this text, we have consciously sought a diverse mix of sources. Each myth is supported by scholarly sources, but many are also supported by pop culture, including blogs and stories from members of the cultures from which the stories originate. This demonstrates the living nature of these stories. We also attempted to use both sources from within cultures as well as academic accounts of these cultures and myths.

INTRODUCTION

All peoples and cultures have narratives interpreting and describing their worldview of humans to others (unknown, supernatural, etc.) Interpretation of myths tells us much about the individual cultural systems.

Dragon myths are much more than stories for children. They represent ancient origins, creation stories, cautionary tales, and explanations of the unknown. Throughout time, the myths are retold, often incorporating cultural, political, religious and natural changes into the cultural systems. Battles between good and evil. Legends of great floods cleansing the earth. Stories of the world being born and prophecies of its demise. Tales of what causes the seasons to change. All myths are born from a human perspective, and many are conceived from the desire to understand and explain the world around us. So, while the prevalence of dragon myths across world cultures may seem outlandish at first, consider the common narrator. Humans have always been witness to the real power of nature. The strength of the elephant, the deadliness of the viper, the endurance of the serpent; these creatures inspire thoughts of even greater beasts—what else might still be out there?

In many of the oldest myths, the most fearsome of these imagined beasts is the great serpent. In nature, snakes possess a number of formidable traits; they can be venomous, powerful, cryptic in coloring, and seemingly capable of vanishing into the earth. Their apparent supernatural abilities and decidedly inhuman appearance frequently earn snakes reverence and prominence in ancient religion. In mythology, these serpentine gods often possess supernatural power and proportions of cosmic scale. Ti'amat, the primordial creation goddess of Mesopotamian origin, is described as a great serpent whose slain body made up the heavens and the earth. The Rainbow Serpents of Australia were said to have shaped the mountains and rivers of the world with their writhing. The Leviathan described in the Bible is said to appear like a serpent or a great fish, and would swallow the souls of the damned. The Four Dragons of China create the four largest rivers in China. It was not uncommon for later translations to call these creatures dragons, for it was the best word available to describe beasts of such strength and power.

The origin of the word "dragon" is complicated but revealing. In the Proto-European language family, "derk" means "to see." In Ancient Greek, "drak", the stem of "derkesthai" means "to see clearly". Then, in Greek, "drakon/draknotos" means "Serpent, Giant Shellfish". In Latin, "draconem" means "Huge serpent, dragon" and in Old French, "dragon" appears as its own term. The evolution of spellings and meanings is fodder for linguists to investigate. Here, we can ponder the question of developing from "to see" to a serpent to a dragon. The Ouroboros seems relevant to this discussion as it is to this entire book. (This will be discussed further with the ever-present draconic symbol of the Ouroboros, who appears again and again throughout this text.)

The modern definition of dragon has of course changed. Eurocentric cultures identify dragons with gold-hoarding and fire breathing, whereas Asiatic cultures see them as water-guardians, keepers of knowledge and ancestors of emperors. Quetzalcoatl, the South-American creation god, is depicted as a glittering green serpent with the wings of a bird and the face of a crocodile. Is he a 'true' dragon? If not, how do we categorize such a being? In North American myth, the Horned Serpents appear again and again as guardian spirits associated with water sources much like Chinese dragons—but rarely are they shown with legs. Are we to deny that these are dragons? There are many amazing myths, comparisons, and interpretations that may be lost should we define dragons as a number of legs or wings. Dragons are much more than this.

In this text, we choose to define dragons as any variation upon the great serpent mythologies. In other words, a supernaturally powerful creature of serpentine form. These great serpents predate our modern definitions of dragon and almost certainly inspired their creation. We posit that it is from these ancient serpents that the more modern, limbed and winged dragons, evolved. They are the definition against which we measure potential dragon myths. In this way, we hope to challenge the reader to redefine the term dragon for themselves, and come to understand more fully the relationship between dragon myths and human culture.

There exist countless myths and variations of myths which involve dragons throughout the world. As it would not be feasible to cover all dragon myths, this text aims to focus its descriptive efforts on those myths with pervasive cultural significance like creation stories or tales that explain the world around us within specific cultural systems. We will discuss the myth itself, along with its biological background, cultural context, and historical background. Prepare yourself; here be dragons.

CHAPTER 1: AFRICA

Introduction:

Myths are born from the human need to explain and understand the unknown. Different cultures have encountered different natural phenomena, political changes and cultural evolution at different times. Hominins, as a species, according to the the theory of evolution, began in Africa. Hence, this book begins the exploration of the pervasive, powerful symbol of dragons on the continent of human origins.

Ouroboros

Background:

One classification of these mythic serpents is the Ouroboros. At its core, an Ouroboros is defined as a serpent or dragon forming a circle by consuming its own tail. The word 'Ouroboros' comes from the Greek words 'oura,' meaning 'tail,' and 'boros,' meaning 'eating;' which produces 'he who eats the tail.' This act is a symbol of balance, the cyclical renewal of life, infinity, and derivatives thereof. Occasionally two or more serpents are involved in the symbol, and in other cases the Ouroboros is not a circle, but a more complex shape. Due to its symbolic meaning, the Ouroboros is commonly featured in myths that explain the cycles of nature and the movement of the heavens.

Since these are popular concepts in world mythology, one can imagine how often the Ouroboros symbol appears. Like the symbol of dragons in general, the Ouroboros symbol touches all corners of the globe and seems to have arisen independently in different locations. The variations of this myth are so plentiful and diverse that they will be described separately for each region of the world. Refer to the two largest dragons in the cover illustration of this book for an interpretation of the two-serpent Ouroboros.

Mehen

Background:

The figure of Mehen can be considered one of the earliest appearances of the Ouroboros, predating Byzantine, Arabic, Jewish and early modern European images of the encircled serpent. Mehen appears throughout ancient Egyptian history, in the Old Kingdom *Pyramid Texts*, the Middle Kingdom *Coffin Texts*, and the New Kingdom *Book(s) of the Dead*. Here, the New Kingdom version of the cycle of day and night is described.

The circle was a common theme in Egyptian religious rites, and could be seen as either a source of protection and order or entrapment, based on the intent of the person using such magic. Mehen's New Kingdom association with Ra's own circular journey around the world, and Mehen's earlier Old Kingdom association with a board game in which players traveled along the serpent's coils to achieve immortality, reflect this association between circular figures and the concept of eternity.

Egyptian culture is known for its focus on the continuance of life, even though many think that the large numbers of tombs reflects a focus on death. But the focus was not death, it was on life everlasting; hence, the need to create tombs and other structures dedicated to life after the life here on earth. Based on agriculture supported by the flooding of the Nile, Egypt was one of the earliest complex cultures, developing many areas of specialization in production. Gender roles in ancient Egypt were far more egalitarian than most of the neighboring Near Eastern cultures. Both men and women could be pharaohs and/or in charge of the sacred temples. Women, upper and lower classes, could work alongside men in various industries and often owned their households. Sexuality, while monogamous during marriages, was also much less constrained for unmarried persons and children were highly valued. Mehen has been referred to as male in some texts, and female in others.

Myth:

In Egyptian mythology, two of the most prominent figures are Ra, god of the Sun, and Osirus, god of darkness. Each day, Ra rises over the earth and brings the life-giving rays of the sun to all. But by night he must travel through the chaotic waters of the underworld by boat, besieged by his mortal enemy Osirus/ Apep. Thankfully, Ra is guarded during this nightly struggle by the protective serpent goddess Mehen. Every night Mehen coils around Ra and deflects the many attacks from Osirus and his ilk. Mehen is almost always successful in her task. On the few days that she fails, the solar eclipse is observed.

In Egyptian art, Mehen is often depicted arced around Ra, clasping her own tail in her mouth. Other works show Ra and Osirus themselves as a two-serpent Ouroboros, locked in their perpetual day and night cycle for all of eternity.

Illustration:

This version of Mehen is based on the cobra (genus: *Naja*). Cobras, unlike pythons and boas, are highly venomous. They are common in ancient Egyptian art regarding serpentine gods. The wings drawn on the hood represent Mehen's protective nature and relation to Ra, god of the sun and sky.

SOURCES:

Aldington, Richard and Delano Ames, Transl. (1959.) New Larousse Encyclopedia of Mythology. Hong Kong: Prometheus Press.

Egyptian Culture. (2013, August 1.) Ancient History Encyclopedia. Retrieved from http://www.ancient.eu/Egyptian_Culture/.

Bane, Theresa. (2015.) Encyclopedia of Beasts and Monsters in Myth, Legend, and Folklore. Jefferson, North Carolina, USA: McFarland & Company, Inc.

Dekirk, Ash. (2006.) Dragonlore: From the Archives of the Grey School of Wizardry. New Jersey: Career Press, Inc.

Kendall, Timothy. (2015, March 29). III. G. Jebel Barkal in the Book of the Dead. Black Drago. Retrieved from http://www. blackdrago.com/fame/mehen.htm.

Roblee, Mark. (2018.) Performing Circles in Ancient Egypt From Mehen to Ouroboros. Preternature: Critical and Historical Studies on the Preternatural, 7(2), 133-153.

Carthaginian Serpent

Background:

Roman history still tells about the summer of 256 B.C./B.C.E when the Roman army decided to invade North Africa. This took place in the midst of the First Punic War, one of many eras of Roman expansion.

Myth:

Consul Marcus Atilius Regulus was leading his vast army into Carthaginian territory, when he eventually had to come to a stop along the banks of the Bagrada River. Carthage was founded as a Phoenician colony, competitors with Rome, by Princess Elissa (Dido) from Tyre.

There, along the river bank, Regulus's armed forces pitched camp for the night. Several soldiers were sent down to the river to gather water, but they soon realized that they were not alone. A serpent of massive proportions rose up, struck and devoured the Roman soldiers.

This prompted the rest of the army to organize and react quickly. Many javelins and darts were hurled at the beast, but to no effect. Its scales formed a curved lattice which sloughed off all potential blows. The serpent was able to move quickly, from one side to the other, with sinuous movement. The articulation between its ribs and belly scales allowed it to mount any terrain and trample over the army easily. As if this wasn't enough, the serpent could bite and then exude a very noxious breath over its attackers.

Regulus considered the situation carefully and decided to besiege the serpent as if it were a true fortress. He ordered the ballistae to be brought up to the river. Then he began hurling large stones at the monster. "A stone taken from a wall was hurled by a ballista; this struck the spine of the serpent and weakened the constitution of its entire body," Orosius (circa 417 A.D./C.E.). With its flexible spine crippled, the serpent's movements were hindered greatly and it was soon overcome by the spears of Regulus's army. They hauled the monster ashore and determined its length to be 120 feet. The Romans then carved off the skin and jaws of the beast. They sent these pieces of the beast back to Rome where they were preserved as a spectacle in a temple. Unfortunately, they were lost around the time of the Numantine War and not seen since.

With no hard evidence, it is difficult to say how large the serpent truly was, but many accounts confirm that some piece of the skin was sent back to Rome as a trophy. It is likely that the Roman army encountered a snake large enough and fearsome enough to warrant killing it and sending it home. What kind of beast was it? We may never know. But, some say this bizarre encounter foretold the defeat of Regulus the following year.

Illustration:

The African rock python (*Python sebae*) is one of the largest living snakes. Native to sub-Saharan Africa, the African rock python can reach lengths of over 15 feet. These pythons are non-venomous, but can constrict and eat large prey such as antelope and goats. They are known to be aggressive if provoked. It may have been possible that African rock pythons once held a range that included Carthage and encountered Roman soldiers. However, while historical specimens theoretically may have been slightly larger, it is unlikely that such a monstrous specimen ever truly existed. It is feasible that the soldiers counted 120 ribs, a feasible number, on the slain snake and this was later mistranslated into feet.

SOURCES:

Bane, Theresa. (2015.) Encyclopedia of Beasts and Monsters in Myth, Legend, and Folklore. Jefferson, North Carolina, USA: McFarland & Company, Inc.

Davis, Nathan. (1861.) Carthage and Her Remains: Being an Account of the Excavations and Researches on the Site of the Phoenician Metropolis in Africa, and Other Adjacent Places. New York, USA: Harper.

Dekirk, Ash. (2006.) Dragonlore: From the Archives of the Grey School of Wizardry. New Jersey: Career Press, Inc.

Stothers, Richard B. (2004.) Ancient Scientific Basis of the 'Great Serpent' from Historical Evidence. Isis 95(2), 220-238.

Grootslang

Background:

In southern Africa, a myth continues about the legendary bottomless pit known as the Wondergat and the monster Grootslang (or Grote Slang for big snake) that lies within. This pit lies deep within the rocky desert of Richtersveld, South Africa and is said to connect to the sea 40 miles away. It is a cave supposedly filled with diamonds and great riches, yet no one dares enter for fear of its legendary guardian, the Grootslang. It is in the territory of the Nama people who are a pastoralist people with patrilineal clans. It is noticeable that the most modern telling of the myth seems to be an anti-colonialist use of the Grootslang as a means to ward off exploitative visitors to the area.

Myth:

The Grootslang is as old as the world itself. In the beginning, the creator gods were still learning how to shape the earth and its creatures. By accident, they made a species with immense strength, deviousness, and intellect. These were the first Grootslangs. Described as huge, snake-like creatures, the Grootslangs were destructive and dangerous by nature, causing trouble wherever they roamed. The gods soon realized that they had made a mistake in the creation of such a powerful creature, and split the Grootslang race into two lesser animals—the elephant and the snake. In the gods' rush to fix their error, however, the greatest and most cunning of the Grootslangs escaped and was never split apart.

Due to this Grootslang's love of diamonds and other treasures, he took up residence in the Wondergat cave in south Africa, home of the Nama people, a pastoral sect of the Khoikhoii culture. The Nama noticed that elephants began to disappear around the bottomless pit. Strange tracks were seen by the nearby Orange river, and there were claims of people seeing a giant snake over 50 feet long roaming the area. There were even tales of the Grootslang transforming into a beautiful maiden and leading men to their deaths in the river. Most knew enough to avoid the Wondergat, but its treasures often lured people to their deaths. Only a few people are known to have entered the cave and survived. When those few emerged, they were radically changed. They had rapidly aged and now had gray hair and terrified eyes. The survivors described the Grootslang as a great serpent with two huge gemstones embedded in its eye sockets and a breath strong enough to knock over a grown man.

One of the most famous Grootslang encounters occurred in 1917. Peter Grayson, an Oxford graduate and successful English businessman, decided to pursue the famous treasure hidden in Richtersveld. "I am determined to return to England as a very rich man or a dead man," he reportedly said before setting sail with six companions in the summer of 1917 (Floyd, 1998). Grayson thoroughly studied the cultures, languages, and topography of the region, determined to ensure a successful venture. Unfortunately, this was not to be the case. On the first night in Richtersveld, one of his companions was killed by a lion. A few nights later, another was killed by a snake bite. After a third fell ill, the party demanded to return to England. Grayson, angered, shouted that he would claim the riches for himself, and abandoned his friends and guide. He was never seen or heard from again. Even today, guides are reluctant to lead expeditions near the Wondergat, and perhaps that is for the best.

Illustration:

Sauropod dinosaurs were huge, herbivorous reptiles that existed during the Mesozoic era. These dinosaurs were characterized by their elongate necks and tails and thick, elephant-like legs and torso. Coming across a fossil of one of these animals, it may seem as though one has discovered the bones of a monster. The long neck and tail appear superficially snake-like, and the appendicular skeleton and ribs are reminiscent of those of an elephant. It is unlikely that a lineage of sauropods survived the Cretaceous-Tertiary extinction event 66 million years ago, but bones of these creatures still exist to this day as fossils. These bones may have inspired myths such as the Grootslang.

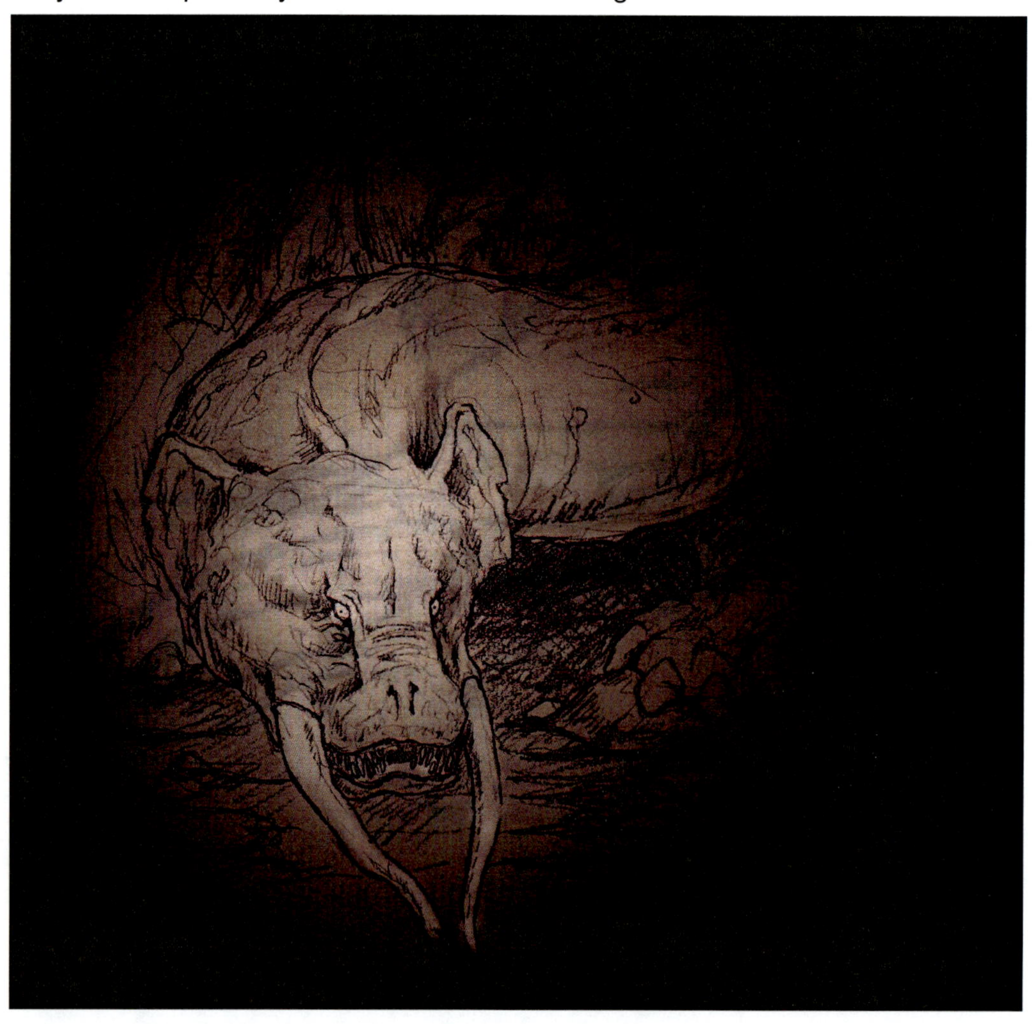

SOURCES:

Bane, Theresa. (2015.) Encyclopedia of Beasts and Monsters in Myth, Legend, and Folklore. Jefferson, North Carolina, USA: McFarland & Company, Inc.

Floyd, Randall. (1998, September 20.) Disappearances feed Grootslang legend. Augusta Chronicle, Retrieved from http://old.chronicle.augusta.com/stories/1998/09/20/ent_239491.shtml.

Richtersveld Route. (2015.) Tourism Route North South. Retrieved from http://www.south-north. co.za/rich_rt.html.

Rose, Carol. (2000.) Giants, Monsters, and Dragons: An Encyclopedia of Folklore, Legend, and Myth. New York, USA: W. W. Norton & Company, Inc.

Ouagadou-Bida

Background:

Somewhere back in the earliest Medieval centuries, the Soninke people settled in the area of the upper Niger River where Mali and Mauritania now are. There they controlled an oasis on a very important east-west trade route through Africa. The toll to pass through this oasis accumulated and became the foundation for the wealth of the medieval kingdom of Ghana. Arab traders called it the land of gold. The Soninke culture was highly stratified, patrilineal, and with an economy based on agriculture (millet is a main crop) and trade.

Myth:

The myth states that when Wagadu (or Ouagadou), the city of the Soninke was first formed, a forest with a sacred grove emerged that surrounded the city and its fresh water. The priest class was responsible for protecting the grove.

At the same time, a great dragon also emerged and lived in a cave within the sacred grove within the forest. This dragon considered it its responsibility to protect the city and the kingdom. The dragon was Ouagadou-Bida. Bida agreed to protect this newly developed city and bestow great fortune on the kingdom and its people if they agreed to send him their most beautiful girl each year as tribute. The people agreed and this was the beginning of great prosperity for the people and kingdom of Ghana.

However, after many generations of sacrificing beautiful young women, a young woman, Sia, was chosen. Sia was in love with the warrior, Amadou Se-fedokote (sometimes written Mamadi Sefe Dekote), and he with her. Amadou was a member of the highly trained warrior class. The couple planned to be married. Amadou did not want to lose his great love to the dragon, not even for the welfare of the rest of the people, the city and the kingdom of Ghana.

One day, close to the time of the sacrifice, Amadou chose to hide near the grove and when the time was right, he emerged and attacked Bida and cut off its head. But Bida did not die. Some say that the head flew off to another city and dropped wealth there. Bida grew another head. It flew off too. Amadou chopped off the second head, then the third, then the fourth, then the fifth and the sixth and the seventh. It was a great bloody battle. Finally, Bida grew no more heads and lay dead. Amadou seized the moment, rescued Sia and together they quickly fled the country. While most tellings give the couple a happy ending, some say that in spite of all his courage, Sia did not love Amadou, and through trickery convinced him to cut off a finger and a toe so that she could reject him for that reason. In turn, he tricked her into sleeping with a servant in his household, and she later died of shame.

Meanwhile, all of Ghana wept because without the protection of Bida, the kingdom no longer prospered. In fact, a great drought overcame the land. Without the protection of the dragon, there was not adequate water and the people fled, taking whatever they could carry with them to survive. This, then, was the end of the medieval kingdom of Ghana.

Illustration:

This image of Ouagado-Bida is based on the Kenyan sand boa (*Eryx muelleri*). This species is native to the southern Sahara Desert. Sand boas' eyes are positioned on the top of the head so they can see above ground when buried in the sand.

SOURCES:

Beachum, L. and Horsky, F. (1971.) African Americans. Grade Eight, Unit Three, 8.3 A & B. Comprehensive Social Studies Curriculum for the Inner City. Report. Youngstown Board of Education, OH.Office of Education (DHEW), Washington, DC.

Blake, Polenth. (2011.) The Dragon Stone-Dragons of Mythology and Fantasy. Retrieved from http://www.polenth.com.

Dekirk, Ash. (2006.) Dragonlore: From the Archives of the Grey School of Wizardry. New Jersey: Career Press, Inc.

Page, Willie F., and R. Hunt Davis, eds. (2005.) "Ougadou-Bida." In Encyclopedia of African History and Culture: African Kingdoms (500 to 1500), vol. 2. New York, USA: Facts on File, Inc.

Nyami Ya Mninga

Background:

The Zambezi River has long been the source of sustenance, physically and spiritually for the Ba Tonga people of southern Zambia, an egalitarian society without the usual social stratification of its neighboring societies. The Ba Tonga settled on both sides of the Zambezi River near the Kariba Gorge. Ba Tonga people follow a matrilineal line of descent and tend to be matriarchal alongside the authority of the priests. The figure of Nyami Ya Mninga is still used by Indigenous leaders today, in both Zambia and neighboring Zimbabwe, as a political message opposing overdevelopment and disregard for traditional cultures. The Zimbabwe Tourism Authority and many travel companies are equally invested in Nyami as a centerpiece to tourism and economic development in the Kariba area.

Myth:

Nyami Ya Mninga (Nyami Nyami is the post-colonization form of the name), the River God, and his wife, protect the Ba Tonga people. Nyami Ya Mninga looks like a very large serpent with the head of a fish or a dragon. Stories say that in times of starvation, Nyami Ya Mninga would come out of the river so people could cut pieces of meat from his body in order to survive. Elders regularly perform ceremonies to honor this very, very large Dragon serpent.

The myths about Nyami Ya Mninga and his wife continue to this day. One of the most recent reaffirms the Ba Tonga belief in their River Dragon God and Goddess.

Colonization and development laid a heavy hand on the land of the Ba Tonga in 1956 in the form of plans to build a dam on the Zambezi River near the Kariwa gorge. Nyami Ya Mninga's wife had gone down river in response to ceremonies of her people. It was while she was gone, in 1957, that dam construction was rapidly moving along. "The worst floods ever known on the Zambezi washed away much of the partly built dam and the heavy equipment, killing many of the workers." (VictoriaFalls24.com).

Some of the bodies mysteriously disappeared. When the relatives of the workers were due to come and retrieve the bodies, dam authorities asked Ba Tonga Elders for aid. The Elders said a sacrifice to Nyami Ya Mninga was necessary to appease the Dragon God and gain his aid. The authorities were skeptical but desperate, so a calf was slaughtered and sent into the river. The next morning, the calf was gone and the bodies of the workers reappeared.

The dam authorities say that crocodiles ate the calf, but have no explanation for the return of the bodies of the workers three days after the flood.

Construction of the dam was completed. The Ba Tonga were forced to relocate higher up on the banks of the Zambezi River. Since the completion of the construction with its very large reservoir of water, there has been increased seismic activity. Scientists say it is due to the weight of the water. Ba Tonga say it is Nyami Ya Mninga, frustrated, trying to reunite with his wife. Ba Tonga say the Dragon God will one day destroy the dam, be reunited with his wife, and its Ba Tonga people will resettle on their ancestral Homeland.

Illustration:

Here, Nyami Nyami is drawn using elements of the tigerfish (genus: *Hydrocynus*) and the bichir (family: Polypteridae). Tigerfish are large, aggressive predators native to freshwater rivers and lakes in Africa. The bichir, another African resident, gets its nickname 'dragon fish' from the many winglike fins on its back.

SOURCES:

Aldridge, Sally. (1978.) The Peoples of Zambia. London, England: Heineman Educational Books.

News24. (2015, October 28.) Angry river god blamed for parched Kariba. Retrieved from https://www.news24.com/Africa/Zimbabwe/Angry-river-god-blamed-for-parched-Kariba-20151028.

Nyaminyami, River God and Spirit of the Zambezi River, Zimbabwe. Zambezi Safari & Travel Company. Retrieved from https://www.zambezi.com/blog/2013/nyaminyami-spirit-zambezi-river/.

Nyaminyami – the Kariba Legend. Zambezi Safari & Travel Company. Retrieved from https://www.zambezi.com/ blog/2011/nyaminyami-the-kariba-legend/.

Nyamukondiwa, Walter. (2018, March 2.) ZTA to Launch Nyaminyami Festival. All Africa, Retrieved from https://allafrica. com/stories/201803020581.html.

The curse of Nyaminyami. (2003, December) Faces: People, Places, and Cultures. Retrieved from http://link.galegroup.com/apps/doc/A112538912/ITOF?u=mlin_w_umassamh&sid=ITOF&xid=f69395ba.

The Legend of Nyami Nyami. (2014, March 21.) Victoria Falls 24.com. Retrieved from http://victoriafalls24.com/blog/2014/03/21/the-legend-of-nyami-nyami/.

Aido Hwedo

Background:

Both an ouroboros and a rainbow serpent, Aido Hwedo appears in the creation stories of the Fon people of Dahomey. The Dahomey kingdom was a west African kingdom originally formed by the Fon people to oppose the domination of the Yoruba in west Africa. The Dahomey was known for fierce warriors, especially a battalion of female warriors, and involvement in the Atlantic slave trade. Theirs was a stratified society of royalty, commoners and slaves with patrilineal descent. Today it is known as Benin.

Myth:

The myth of Aido Hwedo states that Mawu-Lisa, male and female creator twins, made the world with Aido Hwedo as their assistant. At times Aido Hwedo, who is also known as Da, carried them in his mouth. Some say that Aido-Hwedo was created before humankind, some say he was created alongside humankind, and some say that he existed even before Mawu-Lisa. His many movements across the earth created valleys, hills, chasms, and rivers of earth, and where he left piles of excrement, mountains formed.

After Mawu-Lisa had finished creating the earth, they were worried that it would not be able to hold the weight of everything they had set upon it. They asked Aido Hwedo to coil beneath it, grabbing his own tail to support the earth. Because the effort this took was so great and exhausting, Mawu-Lisa created the oceans to keep Aido Hwedo cool and ease his pain. He remains deep in the water, hating heat, but still writhes in agony when the burden of the earth becomes too great. Whenever he moves, there are tremors and earthquakes.

While underwater, Aido Hwedo eats huge quantities of iron. Some say small red monkeys bring this to him. Many say there is a finite store of iron underneath the earth. All agree that once the iron runs out,

Aido Hwedo will not only grab hold of his tail again but in his great hunger, he will start to eat his tail and once his tail is gone, the earth will collapse, driven into the sea by violent earthquakes.

*Note: Mawu-Lisa is sometimes called "he," sometimes "she," and in some variations of this myth a singular male god created the earth with Aido Hwedo at his side. Because most stories featured Mawu-Lisa, and traditionally Mawu is female while Lisa is male - two entities who often operate as one - we chose to use the less gendered, traditionally plural pronoun "they."

Illustration:

Here, Aido Hwedo draws upon general boa anatomy. He has a stout, muscular body suited for holding up the weight of the world. His stripes run longitudinally down his body, like a living rainbow.

SOURCES:

Courlander, Harold. (1996.) A Treasury of African Folklore: The Oral Literature, Traditions, Myths, Legends, Epics, Tales, Recollections, Wisdom, Sayings, and Humor of Africa. New York, USA: Marlowe & Company.

Dekirk, Ash. (2006.) Dragonlore: From the Archives of the Grey School of Wizardry. New Jersey: Career Press, Inc.

Jobes, Gertrude. (1962.) Dictionary of Mythology Folklore and Symbols. New York, NY: The Scarecrow Press, Inc.

Lynch, Patricia Ann. (2004.) African Mythology A to Z. New York, USA: Chelsea House Publishers.

McLeish, Kenneth. (1996). Myth: Myths and Legends of the World Explored. New York, NY: Faces On File, Inc.

Parrinder, Geoffrey. (1961.) West African Religion: A Study of the Beliefs and Practices of Akan, Ewe, Yoruba, Ibo, and Kindred Peoples. London: Epworth Press.

Rose, Carol. (2000.) Giants, Monsters, and Dragons: An Encyclopedia of Folklore, Legend, and Myth. New York, USA: W. W. Norton & Company, Inc.

Turner, Patricia and Charles Russell Coulter. (2000.) Encyclopedia of Ancient Deities. London: Routledge.

The Dragon Of Silene Or St. George And The Dragon

Background:

In the early days of Christendom, Christianity stereotyped pagan religions as devil worship. This devil often took the form of a dragon. Stories of dragon-slayers often evolved/morphed into stories of Saints who conquered the Devil Dragon. The story of St. George and the Dragon is one of these stories.

At first, stories of George told of his many trials and tribulations in defense of the newly formed Christian faith. During the many Crusades of European warriors to the Middle East, these stories inspired European knights to fight harder and longer even though George was not European by birth. He was born in Cappadocia, Turkey. Over time, the slaying of a dragon seemed an obvious addition to the wonders St. George performed.

The story of St. George and the dragon shares some features with St. Clement of Metz and the defeat of the Graoully dragon, in that putting a scarf or stole around a dragon can tame it. Some versions of the St. George story also describe him falling and laying in an exhausted state before reentering the fight, similar to Sir Guy's scheme of feigning death when he fought the Longwitton dragon.

Myth:

In Libya, in the early centuries of the Christian era, there was a city-state, some say a kingdom, known as Silena. Near to Silena was a lake, home to a very voracious dragon. The lake and nearby forest did not have enough food to satiate this dragon, so it would occasionally head into the city, spewing poisonous fumes and consuming whatever or whoever it so desired. At first, the local folk of Silena tried to satisfy the dragon's hunger with two sheep a day. However, this soon depleted the supply of sheep to a point where the flock could not be sustained and the people were facing starvation.

The king decided on an alternative strategy: one sheep and one young man or woman would be sacrificed daily. It didn't matter if the young person was from a rich family or a poor one, one each day had to be sacrificed. Finally, it came to the day that the king's daughter was to be sacrificed. Some stories say the town folk gave the king a week to be with his daughter before she was to be sacrificed. True, or not, the day came when the king's daughter was led to the sacrifice area.

The sound of the hooves of St. George's charger announced his arrival. The king's daughter implored George to hurry away before the dragon came, saw George and ate him, too. George would not run away from or cower in front of the Devil Dragon. Instead, he declared that he would fight the monster in the name of Jesus Christ. And when the dragon showed up, that's exactly what George did. A great battle ensued. Some versions of the battle claim that the dragon had eyes in its wings, which, when open, could bedazzle its opponent. True or not, George was not bedazzled and soon pinned the dragon to the ground with his spear.

Then an amazing miracle happened. George told the king's daughter to wrap her scarf around the dragon's neck. Together, the two led the now complacent dragon into the city. Once in the city, the king and his daughter rejoiced at being reunited, but the townsfolk were shivering with fear at the sight of the dragon. George announced that he would slay the dragon if all in the city renounced devil worship and swore belief in the Christian God. The townspeople agreed to this condition, renounced their pagan beliefs and converted to Christianity. Then George slew the dragon.

Illustration:

Many classical renditions of St. George's dragon place decorative false eyes, or eyespots, on its wings. In nature, many prey animals use eyespots to deter potential predators. Here, they are being used ineffectively to paralyze this dragon's would-be prey with fear.

SOURCES:

Bane, Theresa. (2015.) Encyclopedia of Beasts and Monsters in Myth, Legend, and Folklore. Jefferson, North Carolina, USA: McFarland & Company, Inc.

De Voraigne, Jacobus. (1470, 2012.) The Golden Legend or Lives of the Saints. Princeton, NJ, USA: Princeton University Press.

Evans, Jonathan. (2008.) Dragons. London, UK: Apple Press.

Hodges, Margaret. (1984.) Saint George and the Dragon: A Golden Legend. Boston, MA: Little, Brown, and Company.

Shuker, Karl. (1994.) Dragons: A Natural History. New York: Barnes & Noble Books.

Rose, Carol. (2000.) Giants, Monsters, and Dragons: An Encyclopedia of Folklore, Legend, and Myth. New York, USA: W. W. Norton & Company, Inc.

Isa Bere

Background:

The mountains of Fouta Djallon, in the center of the Republic of Guinea, West Africa, are the home region of the Fula people. Some Fula now live in Sudan, Ethiopia, Mali, Senegal, and Nigeria as well. The Fula have long been an agricultural society, though some Fula (or Fulani) are still nomadic pastoralists. Both agriculturalist and pastoralist Fula people have faced issues with drought even in modern times, and various government efforts to assist them with water deliveries have been largely unsuccessful. The myth of Isa Bere, a very large dragon with an enormous stomach who lived in the mountains of Fouta Djallon, is still part of their history.

Myth:

One day, Isa Bere looked at all of the water of the Tinkisso River, a major tributary of the Niger River, and decided it needed a drink. It began to drink and drink, consuming most of the water of the Niger, causing massive drought throughout West Africa.

King Samba was the son of Faraka. Samba's wife, Queen Annalia Tu-Bari pleaded with Samba to rid the kingdom of the dragon. Samba took his renowned bard, Tarafe, with him to confront this dragon of monumental proportions. For eight long years they fought a terrible bloody battle. Isa Bere broke 800 spears and 80 swords in the process. At last, however, its chest was pierced by Samba's blade, a blade that was specially made for him by the youngest son of a blacksmith of his kingdom. When the dragon was finally split with the sword, the water of the Niger was released from Isa Bere and the kingdom was saved.

Illustration:

Isa Bere is rendered here using elements from the Nile Crocodile (Crocodylus niloticus). Nile Crocodiles are usually 16 feet long but some have been recorded over 20 feet in length. They are known for their strength and aggression.

SOURCES:

Bane, Theresa. (2015.) Encyclopedia of Beasts and Monsters in Myth, Legend, and Folklore. Jefferson, North Carolina, USA: McFarland & Company, Inc.

Danver, Steven L. (2013.) Native Peoples of the World: An Encyclopedia of Groups, Cultures and Contemporary Issues. New York, NY: Routledge.

HN. (2014, November 24.) Soutien d'un Français d'origine africaine. Courrier des lecteurs. Retrieved from https://translate.google.com/translate?hl=en&sl=fr&tl=en&u=http%3A%2F%2Fwww.egaliteetreconciliation.fr%2FCourrier-des-lecteurs-29304.html.

Norwich Dragon Festival. (2014.) Norwich Dragon Festival Education Pack. World Art Collections Exhibitions, Sainsbury Centre for Visual Arts. Retrieved from http://www.heritagecity.org/user_files/downloads/education-pack-vf-web.pdf.

Wikipedia and Livres Groupe. (2011.) Reptile Legendaire : Dragon Legendaire, Naga, Serpent Legendaire, Uraeus, Lindworm, Dragon Oriental, Tarasque, Dragon Occidental, Basilic. Books LLC, Wiki Series.

Cultural Questions:

1. What is your interpretation of the Ougadou Bida legend? Does it represent the rebellion of an underclass (the warriors)? Or trade domination?
2. Water appears in several of the African myths. Compare the importance of water in the different myths
3. St. George slays the dragon and the township must convert to Christianity. What does the dragon represent?
4. Grootslang protects the gems as many dragons seem to do. In this case, what might this represent in the local cultural/economic situation?
5. What is the significance of Aido Hwedo eating iron?
6. We do not know the name of Nyami Ya Mninga's wife. Why do you think this is so?
7. Could the Carthaginian Serpent represent more than a very large serpent, if so, what or who?
8. Discuss the different interpretations of events after the construction of the dam that keeps Nyami na Mninga from his wife.
9. Do any of the myths seem to incorporate themes of colonization, or resistance to colonial ideals?

CHAPTER 2: MIDDLE EAST AND INDIA

Background:

The Middle East and India are regions of the world known for the development of some of the earliest stratified societies as well as early agriculture. The development of irrigation to increase and stabilize agricultural production was necessary in this semi-arid environment, making water an all important resource. A stratified social system accompanied the development of a complex irrigation based agriculture. Access to resources, especially water is at the core of the many wars. War, in fact, is the prime mover for the development of the first political organizations known as states. Water is, unsurprisingly, a consistent theme in many of the dragon myths of this area.

Illuyankas

Background:

Many deities populate ancient Hittite mythology, an agricultural kingdom that rivaled ancient Egypt. The Hittites called themselves Nesili and were a stratified kingdom in the area of contemporary Turkey. They dominated the area from 1700-1200 B.C./B.C.E. (3700-3200 before present). Initially, much of what we know about the Hittites came from their enemies, the Israelites, but later information comes from cylinder seals and art. It is clear that they were a conquering/colonizing power who created an extensive empire with a bilineal kinship system that tended towards patrilineality.

Illuyankas has been compared to Leviathan, which has in turned been compared to Ti'amat. In describing a fight between a water dragon who encompasses chaos and a storm god, the similarities between these dragons seem strong. There are multiple versions of this myth - the most common is presented below. In another version, Illuyankas initially defeats the gods and eats their eyes and hearts, but is ultimately undone by his own daughter who is seduced by Telepinu, son of Taru the weather god. Telepinu convinces her to steal the hearts and eyes from Illuyankas as gifts for him, but immediately gives them back to the gods so that they can defeat Illuyankas and restore order to the universe; this same series of events takes place in the version below, but with Storm God Teshub's son in the place of Telepinu.

Myth:

One of the strongest and most praised deities was the Storm God, Teshub. The Storm God brought rain, rebirth and springtime bounty to the people. However, it was constantly at war with the great chaos dragon Illuyankas. Illuyankas had many-heads and was closely associated with wintertime, death and the sea. Illuyankas was a very greedy beast whose greatest desire was to spread chaos throughout the domain of the gods.

One of the annual battles between Illuyankas and Teshub took place over the city of Kiškilušša. During one such monumental clash, Illuyankas managed to overcome Teshub, the Storm God and bring him to the brink of death. The Storm God escaped, but greatly feared for the survival of the world with Illuyankas on the loose. Teshub called upon his daughter, the goddess Inara, for aid.

Inara decided to take advantage of Illuyankas' greedy nature and throw an enormous feast with such an abundance of wine and beer that no dragon could resist. She then sought the assistance of the mortal hero Ḫupašiya, a man known for his great strength and courage. Inara concealed Hupašiya within the feast hall and invited Illuyankas and all of his children to the celebration. Inara ensured that the plates and glasses of wine never emptied, and soon the dragons grew bloated and very drunk. With the dragons incapacitated, Hupašiya jumped out of hiding and expertly bound all of the creatures together with rope. Then Inara called upon her father, Teshub. The Storm God came and scattered the dragons into many pieces across the land. With this, the chaos dragon was kept at bay for another year.

In another battle with Teshub, Illuyankas managed to encircle him with his coils and pluck out the Storm God's heart and eyes. Weakened, the Storm God married a poor woman and she bore him a son. Teshub asked his son to marry one of Illuyankas' daughters. The son married Illuyankas's daughter and asked his wife for a dowry of the Storm God's eyes and heart. The son returned his father Teshub's eyes and heart. Then the Storm God began his battle with Illuyankas once more. The son, perhaps from the guilt of what he had done, sided with Illuyankas in this fight. However, this did not end well for the son as both he and Illuyankas perished at the hands of the Storm God, and peace was restored to the lands.

Illuyankas and his annual fight with the Storm God not only represent the eventual victory of spring over winter, but are associated with the ancient Hattian festival of the new year, purulli. It is said that the purulli festival celebrates the Storm God's first defeat of Illuyankas and honors the fertility and rebirth of the land in spring.

Illustration:

Illuyankas is shown here, bloated and incapacitated after a wild draconic bender. His children have yet to suffer the ill effects of all the beer.

SOURCES:

Beckman, Gary. (1982.) The Anatolian Myth of Illuyanka. Journal of the Ancient Near Eastern Society 14, 11-25.

Colavito, Jason. (2014.) Teshub and the Dragon. Jason and the Argonauts through the Ages. Retrieved from http://www. argonauts-book.com/teshub-and-the-dragon.html.

Mark, Joshua J. (2016, January 9.) The Hittites. Retrieved from www.ancient.eu/hittite.

McLeish, Kenneth. (1996). Myth: Myths and Legends of the World Explored. New York, NY: Faces On File, Inc.

Rose, Carol. (2000.) Giants, Monsters, and Dragons: An Encyclopedia of Folklore, Legend, and Myth. New York, USA: W. W. Norton & Company, Inc.

Naga

Background:

Naga is the Sanskrit word for serpent. In Hindu mythology, all nagas (and naginis, a feminine form) were descended from Brahma's granddaughter Kadru and her husband Kasyapa-pra-japati. They were considered servants to Indra, who was associated with rain, and enemies with the eagle-god Garuda who was said to eat them as a favorite food. Nagas have been included in lists of minor gods, or devas.

Over time, its meaning expanded to encompass a group of mythical snake beings in Hinduism, Buddhism, and Jainism, belief systems spanning throughout much of southern Asia. These beings are incredibly diverse. Nagas can take forms ranging from snake to human (or a blend of both), can have multiple heads, and sometimes possess the power of shapeshifting. It is said that the Naga race lives in a rich, underground kingdom which goes by many names. As a group, Nagas are not strictly good or evil. However, they are often powerful and usually hold dominion over sources of water such as rivers and lakes. Some guard doorways and others still protect hordes of treasure. Nagas are still worshipped in southern India.

Since the word Naga has become so prevalent, it covers many myths that we consider dragon myths. Here, we will discuss four Nagas that are especially draconic; Apalala, Ananta, Muchalinda, and Vritra.

SOURCES:

Aldington, Richard and Delano Ames, Transl. (1959.) New Larousse Encyclopedia of Mythology. Hong Kong: Prometheus Press.

McLeish, Kenneth. (1996). Myth: Myths and Legends of the World Explored. New York, NY: Faces On File, Inc.

Williams, George M. (2003). Handbook of Hindu Mythology. Santa Barbara, California: ABC-CLIO, Inc.

Apalala

Background:

Between the mountains within modern-day Pakistan, flows the river Swat, an area with an ancient cultural history including Buddhism, Hindu and Islam . The area was an important crossroads for many invading forces and trade as is apparent in the diversity of the religions still practiced there. Springs are the main source of water, and many temples dedicated to naga are located near springs.

Myth:

In ancient times, the people told of a powerful naga named Apalala. Apalala had been a Brahmin who had saved a village from a severe drought. He didn't think that the village folk thanked him adequately so he became reborn as a naga.

In this region there were terrible storms and flood dragons that often tried to cause trouble. Apalala protected the people from these destructive dragons. His protection ensured that the people always had plentiful, healthy crops—and for this they were grateful. In return for his protection, Apalala was given a yearly tribute of grain for his guardianship.

After the passage of many years, the people grew complacent with their way of life, and Apalala was slowly forgotten and denied tribute. This greatly displeased Apalala,he felt disrespected and grew very angry with the people he had once protected. His anger transformed him into a fearsome dragon with the lower body of a serpent, two clawed arms, and the face of a man. He laid siege upon the land with relentless hail storms and floods. He gorged himself upon the grain harvest of the farmers.

The Buddha, hearing of the plight of the region, traveled to the river Swat where Apalala dwelled. He was accompanied by Vajrapani. He approached the dragon unafraid, and spoke to him. The Buddha filled the entire area with fire except for a very small space directly in front of him. Apalala entered the small space and through the wise words of the Buddha, Apalala saw the error in his ways and converted to Buddhism, vowing never to ravage the crop fields again. Fearing starvation, he asked the Buddha what he should do about the grain offerings. The Buddha told him that he should be allowed a single offering once every twelve years. Apalala thanked Buddha and once again brought good fortune to the people of the region. And still when the river floods every twelve years, an offering of grain is always laid on this shore.

Illustration:

This drawing of Apalala mixes the upper body of modern monitor lizards with the lower portion of a serpent. The tribal mask is all that is left of Apala's human face.

SOURCES:

McCall, Gerrie. (2007.) Dragons: Fearsome Monsters from Myth and Fiction. New York, NY: Tangerine Press.

Vogel, Jean Philippe. (1926.) Indian Serpent-lore: Or, The Nagas in Hindu Legend and Art. New Delhi, India: Asian Educational Services.

Vritra

Background:

In India, the swelling rains of the monsoons are integral for the success of the agriculture that sustains the people. For this reason the rains are revered and placed in high regard. In Hindu mythology, Indra is the bringer of rains, king of the gods, and a great warrior against evil. He often clashes against titanic foes and demons, the most famous of which is his nemesis who goes by the name Vritra. In later versions of the story, particularly in Puranic literature, his powers were reduced in the myth and he required the help of other gods in defeating Vritra.

A later story in the Puranas describes Vritra as having once been human, but having offended Parvati by spying on her while she was with Shiva, he was cursed to be reborn as a draconic water demon. His name translates approximately into “one who holds back.”

This myth shares similarities with Illuyankas and Ti’amat, with Indra as a storm god representing order and Vritra as a water dragon representing chaos.

Myth:

Vritra is a powerful drought demon. He is described as being a serpent so large his head reaches the sky and with one swipe of his tail, he can level mountains. In a display of power, Vritra decided to steal all of the water in the world, sequestering the rivers and rains into a mountain and keeping it all for himself. This act plagued the people of India with severe thirst and dying crops. Many warriors were afraid to challenge the terrible Vritra. However, Indra rose to the challenge. He armed himself with his trusted weapon, the lightning bolt, and imbibed a great deal of the intoxicating ritual drink soma. Indra approached Vritra’s mountain fortress and challenged him to battle. Some say he was riding his elephant steed.

In a fight of epic proportions, Indra charged at the dragon with his lightning bolts. Vritra writhed about and clawed at the storm god. They exchanged powerful hits for a very long time, but both beings were very evenly matched. Eventually, Indra succeeded in severing Vritra’s limbs as the dragon drew too close to him. Vritra gave out a mighty scream. Then Indra landed the killing blow, cleaving the dragon in two and releasing the waters of the world. In some versions of the fight, Vritra’s mother arrived to avenge the death of her son before Indra could recover the waters. That fight was just as challenging and epic, but Indra succeeded in slaying her as well.

Another episode of their ongoing encounters describes a time when Vritra agreed to stop fighting Indra, and that Indra would never attack, neither by day or night nor with a weapon of wood, stone or iron. But, Indra desired revenge and one day while meditating, saw that it was dusk, in between day and night and saw a huge amount of foam heading to shore. Indra used the foam to defeat Vitra without breaking his promise.

A version of this myth is often used to describe the annual coming of the monsoons. In this case, the rains are a result of Indra beating the waters out from the belly of Vritra, and since both are equally matched the fight renews itself again year after year.

Illustration:

The villainous Vritra sits in the above image, the waters of the world seeping out of his great and terrible maw. This Vritra takes inspiration from Indian statues and the bizarre horned frog species of southeast Asia.

SOURCES:

Aldington, Richard and Delano Ames, Transl. (1959.) New Larousse Encyclopedia of Mythology. Hong Kong: Prometheus Press.

Dekirk, Ash. (2006.) Dragonlore: From the Archives of the Grey School of Wizardry. New Jersey: Career Press, Inc.

McLeish, Kenneth. (1996). Myth: Myths and Legends of the World Explored. New York, NY: Faces On File, Inc.

Stookey, Lorena Laura. (2003.) Thematic Guide to World Mythology. Santa Barbara, CA: Greenwood Publishing Group.

Williams, George M. (2003). Handbook of Hindu Mythology. Santa Barbara, California: ABC-CLIO, Inc.

Muchalinda

Background:

Nagas are common in Buddhist mythology, often seen as guardian figures. This is easily explained in the connection between Hindu mythology, the original religion of the Buddha. This myth appears frequently in Buddhist art, and can be seen in relief sculptures in Thailand, Myanmar, Cambodia, Laos, and Sri Lanka, though the meaning of the potent symbol of the naga protecting Buddha may be different.

Myth:

In a famous story, the king of all Nagas, Muchalinda, came across the Buddha meditating under a Bo tree. Some stories suggest that Muchalinda's home was in the roots of the Bo tree. Muchalinda noticed the sky grow dark and smelled the coming of rain and thunder. To protect Buddha from the storm, which raged for seven days, Muchalinda transformed into an enormous hooded cobra and wound himself seven times around the Bo tree. His hood shielded Buddha from the storm and he was able to continue to meditate in safety. In some stories, after the storm passed Muchalinda transformed into a young man and bowed to Buddha, becoming the first creature to recognize him.

Illustration:

Muchalinda is the most powerful of the nagas, and is usually seen with 7 cobra heads with their hoods extended. This drawing of Muchalinda was heavily influenced by traditional statues involving multi-headed nagas and the Indian cobra Naja naja. The genus name Naja is derived from the Sanskrit word Naga, meaning snake.

SOURCES:

McLeish, Kenneth. (1996). Myth: Myths and Legends of the World Explored. New York, NY: Faces On File, Inc.

Sharrock, Peter D. (2015.) Serpent-enthroned Buddha of Angkor. Marg, A Magazine of the Arts, 20.

Ananta Sesha

Background:

According to the Mahabharata, Ananta Sesha was the first of all the Nagas. He existed before the gods and the world itself. Ananta roughly translates to "the endless one," and Sesha, "all that is left," for he is believed to both cause and survive the end of the world. He is described as a glowing white serpent of unknowable proportions, inhabiting the great cosmic ocean before creation. Some sources describe him as the king of all nagas; others say that he copied himself to create every snake on earth.

Myth:

According to one story, the supreme protector deity Vishnu reclined upon the endless coils of Ananta when he first conceived the world. Ananta is described as possessing either 7 or 1000 heads, each hooded and capable of spewing poison, fire, and in some cases, perfume. In some versions of the story, Ananta's horns and body are studded with gems so bright they can illuminate the underworld.

Following the creation of the world and the birth of the naga race, Ananta Sesha ruled over the subterranean Naga kingdom. During an argument between Kadru (the mother of all nagas) and her sister, many of the naga were cursed by Kadru for refusing to aid her in cheating during a bet. Following this, Ananta decided to remove himself from his brethren and practice penance to mourn the situation. The creator god, Brahma, saw Ananta's suffering and gave him counsel. Ananta wished to continue practicing pious austerity, so Brahma suggested that Ananta could continue his pious ways and that this would support the earth and keep it steady. Ananta Sesha agreed to this and carefully placed the world atop his head like a crown. According to the myth, Ananta is destined to destroy the world at the end of each age by consuming it with fire.

Illustration:

Here Ananta Sesha is shown bathing in the cosmic ocean with Vishnu reclining on his back. His coils curl infinitely through the clouds with no end. Again, like the other nagas, his design is inspired by the spectacled cobra, Naja naja, native to India.

SOURCES:

Aldington, Richard and Delano Ames, Transl. (1959.) New Larousse Encyclopedia of Mythology. Hong Kong: Prometheus Press.

Allen, Judy and Jeanne Griffiths. (1979.) The Book of the Dragon. New Jersey: Garnstone Press.

Dekirk, Ash. (2006.) Dragonlore: From the Archives of the Grey School of Wizardry. New Jersey: Career Press, Inc.

McCormick, Kylie. (2017, October 9). Ananta. Dragons of Fame. Retrieved from http://www.blackdrago.com/ fame/ananta.htm.

McLeish, Kenneth. (1996). Myth: Myths and Legends of the World Explored. New York, NY: Faces On File, Inc.

Niles, Doug. (2013.) Dragons: the Myths, Legends, & Lore. Avon, Massachusetts: Adams Media.

Rose, Carol. (2000.) Giants, Monsters, and Dragons: An Encyclopedia of Folklore, Legend, and Myth. New York, USA: W. W. Norton & Company, Inc.

Ti'amat

Background:

"In the beginning" is how many creation myths are told. However, sometimes the myths change through time, especially when the dominant political forces change. Often it is said that history is written by the conquerors. The myth of T'iamat is an example of the forces of political change and domination, altering the content of a myth. Ti'amat was one of the creation gods of Sumer, which was a major political force in the Middle East around 4000-5000 B.C./B.C.E. Babylon was dominant around 2500 B.C./B.C.E.

While written versions of the myth largely come from the Babylonian epic of creation, the Enuma Elish, and later iterations of the story, the original is likely much older. The Khafaje clay plaque from the Isin-Larsa time period (2004-1763 B.C./B.C.E.), for instance, predates the heyday of Babylon but visually depicts a clear case of the Ti'amat myth. While some scholars believe that stories of storm gods fighting water dragons emerged in the Mediterranean where the environment could have more easily inspired such imagery, others insist that the T'iamat story emerged in the Near East and became a major influence to tales such as Illuyankas, Leviathan, and Biblical serpentine imagery. Later Babylonian stories foretold that Ti'amat may return from the dead to help end the earth, should Marduk ever stand up and disrupt the balance of earth, which sits on his shoulders.

Myth:

The Babylonian creation myth that has been told for centuries, possibly millennia, exalts Marduk, in particular, as the great warrior, bringer of order, who defeats the evil dragon, Ti'amat, who is chaos personified.

But in the earlier myth, told in Sumer, the beginning is different.

Nothing existed, not a tree, house or a city. The world was seen as a vast sea. It is Ti'amat, a primordial being, and Apsu who join together to create the world. Ti'amat, the goddess of the salt seas and Apsu, the deity of fresh waters create the beginning of the world. They bring numerous children into the world, lesser gods. In fact, they had many children who are described as gods and goddesses themselves. Their children were Mummu, Lahmu and Lahamu (twins who were otherwise known as Ea and Damkina), Anshar, and Kishar. Mummu was a close advisor to the parents.

Ea and Damkina are the parents of Marduk, a god with control over storms whose name itself means "son of the storm."

Many later versions depict Ti'amat as a vicious vengeful demon, set on destroying the younger gods. But in the earliest translations of the cuneiform texts, Apsu is disturbed nightly by the revelries of the younger gods and wants to disperse them throughout many lands so that he and Ti'amat may once more have restful days and sleepful nights. However, Apsu's plan was overheard and Ea creates his own plan to destroy his parents, Apsu and Ti'amat.

The plan succeeds in destroying Apsu, but not Ti'amat. In her grief, Ti'amat swears revenge and organizes a massive war effort against Ea, Marduk and their associates. Ti'amat makes Kingu, one of her children, in charge of her forces. But the opposing forces choose Marduk as their avenger, and he went forth with spear, bow and arrow, club and lightning.

> *"He made a net to enclose the inward parts of Ti'amat,*
> *The four winds he stationed so that nothing of her might escape;*
> ....
> *He sent forth the winds which he had created, the seven of them;*
> *To disturb the inward parts of Ti'amat, they followed after him.*
> *Then the lord raised the thunderbolt, his mighty weapon,*
> *He mounted the chariot, the storm unequalled for terror,"*
> (Leonard King, The Seven Tablets of Creation)

Ultimately Ti'amat and Marduk face each other in battle, their lesser god allies ready to accept the fate of this mega battle. Marduk reigns supreme and Tia'mat goes down in history as the evil, unreasonable, irrational, force of chaos, defeated by the wondrous god, Marduk, ruler of Babylonia.

And Sumer is conquered.

The body of Ti'amat becomes the earth and all of the features on it, the mountains, the river valleys, even the sky, but in the beginning, she was the mother of all... and some stories say she still is.

Illustration:

Ti'amat floats here in the cosmic abyss, the mountains of the world running down her back. Her design is based loosely on the sturgeon, as seen in the two rows of bony scutes running down her sides. Sturgeons are a primitive group of fishes, much like how Ti'amat is one of the first recorded dragon myths.

SOURCES:

Aldington, Richard and Delano Ames, Transl. (1959.) New Larousse Encyclopedia of Mythology. Hong Kong: Prometheus Press.

Barton, George A. (1893.) Tiamat. Journal of the American Oriental Society 15, 1-27.

Evans, Jonathan. (2008.) Dragons. London, UK: Apple Press.

Grafman, R. (1972.) Bringing Tiamat to Earth. Israel Exploration Journal 22(1), 47-49.

Ingersoll, Ernest. (1928, 2014.) Dragons and Dragon Lore. New York: Cosimo Classics.

Jacobsen, Thorkild. (1977.) Mesopotamia. In Before Philosophy: The Intellectual Adventure of Ancient Man. Chicago, IL: University of Chicago Press.

Jacobsen, Thorkild. (1968.) The Battle between Marduk and Tiamat. Journal of the American Oriental Society 88(1), 104-108.

Jobes, Gertrude. (1962.) Dictionary of Mythology Folklore and Symbols. New York, NY: The Scarecrow Press, Inc.

King, Leonard. (1902, 2014.) The Seven Tablets of Creation. CreateSpace Publishing Platform.

McCall, Henrietta. (1990.) Mesopotamian Myths. London, UK: British Museum Press.

McLeish, Kenneth. (1996). Myth: Myths and Legends of the World Explored. New York, NY: Faces On File, Inc.

Rose, Carol. (2000.) Giants, Monsters, and Dragons: An Encyclopedia of Folklore, Legend, and Myth. New York, USA: W. W. Norton & Company, Inc.

Shuker, Karl. (1994.) Dragons: A Natural History. New York: Barnes & Noble Books.

Leviathan and Behemoth

Background:

Ancient Hebrew texts such as the Old Testament often contain descriptions of great monsters and dragons. Perhaps the most famous of which is the Leviathan. Along with Tiamat, there were many ancient mythologies which contributed to the myth of the great Leviathan. One story tells of the storm god Baal fighting the seven-headed sea monster named lotan, though in the Ras Shamra texts of Ugarit, now Syria, from around 1700 B.C./B.C.E., it is Baal's sister and consort Anat who defeats the sea dragon. In fact, some tales directly name this seven-headed fire-breathing crocodile-like "lotan" as Leviathan, placing its origin prior to the Hebrew and Christian tales of it.

It is believed that the word Leviathan derived from the world lotan, meaning wrapped, or hidden and the Hebrew word 'tehom' means deep. The existence of colossal monsters/dragons such as Leviathan is often seen as evidence of the power of deities.

Myth:

Leviathan is first described by name in Job 41 in the Old Testament. In this section, God describes the power and majesty of his creations, and his dominion over them.

> 1 Can you pull in Leviathan with a fishhook
> or tie down its tongue with a rope?
> 2 Can you put a cord through its nose
> or pierce its jaw with a hook?

God describes Leviathan as untamable and unable to be harmed by any weapon. Its physical description features many draconic elements.

> 14 Who dares open the doors of its mouth,
> ringed about with fearsome teeth?
> 15 Its back has rows of shields
> tightly sealed together;
> ...
> 18 Its snorting throws out flashes of light;
> its eyes are like the rays of dawn.
> 19 Flames stream from its mouth;
> sparks of fire shoot out.
> (Job 41 New International Version)

When Leviathan moves it churns the entire ocean. No other earthly creature is its equal. Although, another creature described in Job 40 is praised in a similar way by God. Where Leviathan is the king of the seas, Behemoth is the greatest of the land creatures. Only its creator, God, was strong enough to lay a hand against it.

> 15 Look at Behemoth,
> which I made along with you
> and which feeds on grass like an ox.
> ...
> 17 Its tail sways like a cedar;
> the sinews of its thighs are close-knit.
> 18 Its bones are tubes of bronze,
> its limbs like rods of iron.

Behemoth is often depicted as either an ox or a hippopotamus, but is usually described alongside the draconic Leviathan. Some Apocrypha describe God creating the male Behemoth for the female Leviathan, or even two Leviathans of differing genders. In these stories the female Leviathan is killed so they cannot reproduce and doom the rest of creation. However, the other is made immortal. The meat of the female is saved for later to be served as a banquet to the righteous on Judgement Day. Still other iterations speak of God slaying both Behemoth and Leviathan when Israel is delivered from all enemies to be served as food for the righteous.

Interestingly, Leviathan takes on a more sinister role in later renditions of Biblical stories. It is rendered as a 'Hellmouth,' swallowing the souls of the damned during the Day of Judgement. Its destruction at the hands of God is sometimes interpreted as God's strength overcoming the forces of evil in the world. There are even similarities between Leviathan's mythological predecessors and another famous Biblical dragon who appears in Revelation: Satan. In Rev. 12:3… "Then another sign appeared in heaven: and behold, a great red dragon having seven heads and ten horns, and on his heads were seven diadems." This dragon fights the forces of heaven and is thrown down to earth with his angels. In Rev. 13:1, the same red dragon is seen coming out of the sea and is the enemy of God. This association between dragons and the forces of evil in the world will play a large role in shaping the dragon myths in later chapters involving Christian Europe.

Illustrations:

Here, Behemoth towers over the mountains like a gargantuan sauropod dinosaur. He is historically depicted as a bull, so he is drawn here with bull-like horns and ears.

The image of Leviathan is based on the mosasaur. Mosasaurs are an extinct group of marine reptiles, distantly related to modern monitor lizards. These animals were amazing aquatic predators, and it is thought that they could swallow their prey whole.

SOURCES:

Dekirk, Ash. (2006.) Dragonlore: From the Archives of the Grey School of Wizardry. New Jersey: Career Press, Inc.

Hirsch, Emil, Kaufmann Kohler, Solomon Schechter, Isaac Broyde. Leviathan and Behemoth. Jewish Encyclopedia. Retrieved from http://www.jewishencyclopedia.com/articles/9841-leviathan-and-behemoth.

McLeish, Kenneth. (1996). Myth: Myths and Legends of the World Explored. New York, NY: Faces On File, Inc.

Revelation 12:3. Bible.hub. Retrieved from http://biblehub.com/revelation/12-3.htm.

Wallace, Howard. (1948.) Leviathan and the Beast in Revelation. The Biblical Archaeologist 11(3), 61-68.

Gandarewa

Background:

From the texts of ancient Zoroastria through to ancient texts of Sumeria, Persia, India, and even into contemporary Iran, stories exist about an immense dragon, Gandarewa (alias Gandareva, Kundra or Kundraw). The complete history of Zoroatrianism is unclear although the religion seems to have originated somewhere east of Iran and is claimed to be the first "world" religion and the oldest religion that is still active. It is a monotheistic religion.

Zoroaster (Zarathustra in ancient Persian) is thought to have lived somewhere between 600 and 1500 B.C./B.C.E., possibly born in northeastern Iran or southwestern Afghanistan. Most of what we know about him comes from Zoroastrian scriptures, the Avesta, although this has now been added to by linguistic and archaeological data. Connections to other world religions include the significance of the number 9 in Sumerian mythology, and several parallels between Gandarewa and the Hindu Gandharva, who battles Indra.

Myth:

Gandarewa is first known as a water dragon, so large that he stretches from the ocean to the sky. His head is often said to be lit by the sun, while his body is hidden to the depths.

In later stories, Gandarewa is seen as malevolent, capable of swallowing many men and horses at the same time. But in the earlier tales, he is described as a benevolent god, and maintains the title "golden-heeled" as a reminder of his kinder days. Ultimately, he was made greedy by his need to guard the haoma, the main healing plant desired by all for its healing abilities. His desire to control the gift of haoma, only allowing those he chose access to it to benefit from its healing abilities, led to his possessiveness and greed. He was also known to collude with Azi Dahaka, often spying for him.

Gandarewa is said to have ruled the deep sea. His greed led to a hunger for the destruction of creation and that developed into the slaughter of both humans and livestock. Many saw Gandarewa as unbeatable, but the great monster slayer of Persian mythology, Keresaspa, saw this carnage and challenged the great dragon to battle.

When Keresaspa approached Gandarewa in the sea, he saw the bones of dead men hanging from the beast's jaws. Gandarewa saw the hero and lunged forward, grabbing his beard, and the fight began. For nine days Keresaspa fought an uphill battle with the enormous dragon. His luck turned when he managed to grasp the skin on the sole of Gandarewa's foot and in one sweeping motion flayed all the skin from his body. Keresaspa quickly bound the dragon with the dragon's own skin as he was screaming in agony. Then Keresaspa dragged the dragon to the shore.

Gandarewa was not defeated so easily. Once ashore, he whipped the eyes of Keresaspa and sent him flailing blindly into the trees. Gandarewa then ate all fifteen of Keresaspa's horses and kidnapped his wife and children. He brought the family to his watery kingdom. Enraged, Keresaspa heroically regained his strength and walked into the ocean and dragged Gandarewa back ashore. With a swift blow from his club to the dragon's skull, Gandarewa was slain and Keresaspa's family and all of Persia was saved.

Illustration:

In this image, Gandarewa guards the White Haoma plant, deep under the waves. This version of Gandarewa draws influence from the Permian fossil amphibian †Diplocaulus. His skin and external gills are like those seen in extant larval salamanders. Of course, real amphibians cannot survive in salt water.

SOURCES:

Carnoy, Albert J. (1917.) Iranian Mythology. Boston : Marshall Jones company, 1917.

Dekirk, Ash. (2006.) Dragonlore: From the Archives of the Grey School of Wizardry. New Jersey: Career Press, Inc.

Jobes, Gertrude. (1962.) Dictionary of Mythology Folklore and Symbols. New York, NY: The Scarecrow Press, Inc.

Kramer, Samuel Noah. (1944, 2007.) Sumerian Mythology: A Study of Spiritual and Literary Achievement in the Third Millennium B.C. London: Forgotten Books.

McCormick, Kylie. (2017, October 9.) Gandarewa. Dragons of Fame. Retrieved from http://www.blackdrago.com/fame/ gandareva.htm.

Zoroastrianism. (2016, April 4). New World Encyclopedia, Retrieved from http://www.newworldencyclopedia.org/p/index.php?title=Zoroastrianism&oldid=995084.

Azi Dahaka

Background:

As mentioned above, Zoroastrian culture and mythology is one of the oldest documented religions. Zoroastra is seen in ancient scripts and other sources as the earliest example of a widespread religious leader who reaches many cultures and becomes a renowned spiritual icon. The mythology that developed around the religion contains stories of the dragon, Azi Dahaka.

There are multiple aspects given to Azi Dahaka over time; in the Avesta, the earliest Zoroastrian text, he is a three-headed, six-eyed dragon with three pairs of fangs who rules the second millennium of human history. Some related myths state that his three heads represent pain, anguish, and death. In the later Bundahesh text, he commits incest with his own mother. In the Shahnamah, an epic of Persian kings, he is a human with two serpents growing from his shoulders. In all of the stories, he is a destroyer who brings drought, storms, cold, disease, and death. Some scholars believe that he may be the personification of the Babylonian empire, which ruled over Iran.

Myth:

Azi Dahaka appears in many forms, the most ancient of which comes from Zoroastrian mythology. There, he is the greatest of the Draj demons with three heads and a thousand senses. He was born to cause destruction and destroy faithful souls. At first, his hunger led him to consume cattle and other livestock, but soon this was not enough to sate him and he began eating humans. He represents lies and evil.

As terror spread through the land, the hero Thraetaona rose up and challenged the dragon. The two engaged in combat for days. Thraetaona landed many hits against Azi's head and heart with his club, but seemingly nothing would stop Azi Dahaka's rage. Thraetaona then raised a sword against the dragon, but wherever the blade penetrated the skin, terrible creatures spilled out of the wounds instead of blood. Fearing that the world might overflow with disgusting animals if the battle continued, Thraetaona figured that imprisonment was the only option left for defeating this beast. The hero backed Azi Dahaka up into a cavern at the top of a mountain and succeeded in chaining him there. The fight was won, but it was said that whenever earthquakes shook the land, they were the result of the dragon attempting to break out of his prison.

It was believed that Azi Dahaka will break free of his bonds before the world's end and will cause great destruction, but will be defeated. In some versions, this hero is Thraetaona; in others, the Zoroastrian hero Keresaspa returns. In yet other tellings, the hero is not Keresaspa but supreme god Ahura Mazda's own son Atar who fights Azi Dahaka and chains him to the mountain to await that final battle.

In later stories, Azi Dahaka returns by possessing an innocent man with his evil spirit. In this tale, there is a naive Iranian prince next in line for the throne. One day, a sorcerer named Ahriman appears before the prince and describes to him all of the luxuries which come with claiming the throne. The prince is greatly tempted by what the sorcerer tells him and is tricked into committing patricide on the king. The prince, beginning to fall victim to Azi Dahaka's evil influence, is further tempted by the sorcerer to indulge in eating meat. The prince does so and falls further under the dragon's influence.

At this point, the sorcerer continues to flatter the prince, and asks if he would be gracious enough to let him kiss the prince's shoulders. The prince agrees, and two snakes grow from where the sorcerer's lips touched him. The prince attempts to remove the snakes, but they grow back whenever cut and become more and more aggressive. The snakes take on Azi Dahaka's insatiable hunger, and must be fed human flesh every day. At this point the prince has been completely possessed by the evil dragon, and he goes on to rule the Persian empire for 1000 years. His reign comes to an end when he has a dream foretelling of a glorious hero that will defeat him. This hero appears with many different names and identities, and sometimes it is the hero Keresaspa (or Kereshapa) who frees the Persian people from their oppressive ruler.

Illustration:

Azi Dahaka has many forms, one of which is the three-headed dragon. This illustration uses features from *Cerastes cerastes*, a venomous viper from Africa and the Middle East. Their cryptic patterning helps them camouflage with the desert environment. The supraorbital horns are natural in this species.

SOURCES:

Aldington, Richard and Delano Ames, Transl. (1959.) New Larousse Encyclopedia of Mythology. Hong Kong: Prometheus Press.

Azdaha. (2011, August 18.) Encyclopaedia Iranica. Retrieved from http://www.iranicaonline.org/articles/azdaha-dragon- various-kinds#pt1.

Dekirk, Ash. (2006.) Dragonlore: From the Archives of the Grey School of Wizardry. New Jersey: Career Press, Inc.

Jobes, Gertrude. (1962.) Dictionary of Mythology Folklore and Symbols. New York, NY: The Scarecrow Press, Inc.

McLeish, Kenneth. (1996). Myth: Myths and Legends of the World Explored. New York, NY: Faces On File, Inc.

Peterson, Joseph H. (1995.) Avesta: Khorda Avesta. Avesta.org. Retrieved from http://www.avesta.org/ka/yt5sbe.htm.

Cultural Questions:

1. In both the myth of Ti'amat and the myth Illuyankas, familial conflict is a central theme. Discuss evidence of familial conflict in ancient Sumer/Babylon and in India in general.
2. Myths reflect their cultural/ecological origins. How does this affect our interpretations of Leviathan/Behemoth? Do our cultural origins impact our interpretations?
3. The Zoroastrian cultural system is the source of two of the myths for this chapter. How do you think the hero, Keresaspa, fits into the Zoroastrian and Persian cultural systems? Could he be the source of a monotheistic religion? Divine right of kings/emperors?
4. What is the significance of the many Naga myths? How do they fit into the Hindu cultural system?
5. Discuss the addition of Buddha to myths of dragons and nagas.

CHAPTER 3: SOUTHEAST ASIA

Introduction:

The Southeast Asian region is home to many diverse cultures in a wide variety of ecological niches. The geographic area includes many, many rivers as well as mountains, seas and hot springs from a heated earth core (think volcanoes). Many of the dragon legends in this area are concerned with the relationship of water and creatures that live in it or control it and their interactions with the humans who live in their region. The area is one of the first human areas to develop stratified societies, and some say that this is reflected in the dragon legends.

Makara

Background:

Hinduism is a complex cultural system based in South Asia. It's estimated to have originated in the Indus Valley or, some say, on the "other side" of the Indus River. It has often been referred to as the eternal religion, possibly the world's oldest religion, and currently the world's third largest. It is definitely one of the most complicated. Cultures which embrace Hinduism have historically been patrilineal and patriarchal, though Hinduism includes many powerful goddesses. These include societies that utilize the caste system, a rigid stratification of society, regulated by marriage rules and assigned occupations. Reincarnation plays a role in both the religion and the caste system.

Myth:

In the Vedic tradition, on the subcontinent of India, the goddess Ganga and the god Varuna claim the dragon Makara as their personal transport when crossing bodies of water. Makara is a Sanskrit word meaning 'sea monster,' but this does not adequately describe Makara.

Interestingly, the Makara is described as possessing aspects from both land and sea creatures. The oldest drawings give Makara the lower jaw of the crocodile, an elephant trunk for its upper jaw, the teeth and ears of a boar, and the eyes of a monkey for the front half of its body. The back half is lithe and scaled like a fish, but interspersed with peacock feathers. Later on, Makara takes on the parts of other animals like a horse, a lion, and a deer, but it is always a land animal in the front and an aquatic animal in the back. The symbol of Makara is the tenth in the Indian zodiac, which corresponds to the well-known sea-goat Capricorn in the western zodiac.

The dragon Makara became associated with strength and tenacity, due to its resemblance to a crocodile. For this reason it was popular to engrave weapons with the likeness of Makara. Makara also became known as a doorway guardian and rows of Makara are known to decorate the base of temples. This, along with its status as a water creature, led to many sculptures of Makara as roof gargoyles that spout rainwater. It is common amongst sculptures in this region to have Makara flanked by two Nagas; dragons also associated with deities and water. Makara are seen in architecture from India to Cambodia.

Illustration:

Makara is drawn here with elements of a crocodile, elephant, and boar in front half of its body and those of a fish and peacock in the back half. He also has a saddle so he can carry the gods across bodies of water.

SOURCES:

Beer, Robert. (2003.) The Handbook of Tibetan Buddhist Symbols. Boston, MA, USA: Shambhala, Pub., Inc.

Dobson, Kenneth and Arthur Saniotis. (2014.) Dragons: Myth and the Cosmic Powers. Prajna Vihara: Journal of Philosophy and Religion 15(1), 85-99.

Krishnamurphy, K. (1984.) Mythical Animals in Indian Art. New Delhi, India: Abhinv Publications.

Williams, George M. (2003). Handbook of Hindu Mythology. Santa Barbara, California: ABC-CLIO, Inc.

Baruklinting

Background:

Java is part of Indonesia and named after the Javanese culture. There is a long history of association with both Hindu and Buddhist traditions. The region is known for its active volcanoes; in fact, much of the Indonesian region was created by volcanoes. "Java Man" is a 1.7 million old hominin form from this region, demonstrating the spread of human ancestors throughout the Southeast. It is an agricultural society, patrilineal and patriarchal, where rice is the major crop and blacksmithing was a high ranked occupation.

In the Rawa Pening area of Central Java, Indonesia, this story is quite different. In that version, Baru Klinting is described as "Naga," and is initially angry at a village that not only denies him food, but chops up his body and eats him at a festival. He regenerates and transforms into a beggar boy, and an old woman takes pity on him. He then pulls a stick from the ground and releases a flood in the village that creates Lake Rawa Pening, but gives a boat to the old woman. This version of the story explains an ecological phenomenon, but also imparts cultural values. It is still used in classrooms in that area and is sometimes simply referred to as "the story of Rawa Pening."

Myth:

There is a famous story hailing from Java about the origin of the village Kesongo. Long ago, the region was inhabited by a massive dragon with a gaping mouth. This was Baruklinting. While meditating, the dragon foresaw that he would deliver 9 shepherds as an offering to the gods. Baruklinting lumbered over to the nearest cow fields and hatched a plan.

These cow fields belonged to a village in which 10 shepherd boys lived. Nine of the boys were strong and fit, but one was skinny, dirty, and covered in ulcers. Unsurprisingly, he was often at the receiving end of pranks and torment. He never fought back, and he would often find the nicest fields for his cows to graze and stay far away from the other shepherds.

One day, the other nine shepherds saw that the ugly shepherd had found the perfect tree to take a nap against. Envious, they plotted to take the dung from their cow fields and fling it at him to make him leave. The ugly shepherd heard their plan, and got ready to run for it, when suddenly the sky opened up with rain. Lightning sliced the sky. All the shepherds scattered, searching for shelter from the storm. The ugly shepherd was smart and soon found a rock-filled cave in which to hide from the rain. The other shepherds saw this and rushed into the cave, kicking the ugly shepherd out and stealing his shelter. As soon as he fell out of the cave, the cave snapped shut on the nine shepherd boys.

The ugly shepherd looked up higher and saw that the cave was not a cave at all, but the mouth of Baruklinting! Shaking with fright, the only remaining shepherd ran back to the village. He told the villagers what had happened, and everyone grabbed weapons in preparation for dragon slaying. But when they returned to where the dragon had sat, there was no trace of him. The rain and lightning had stopped. The dragon and the nine shepherds were no longer there.

Henceforth, the village was always known as Kesongo; songo meaning nine, for the nine selfish shepherds who were eaten by Baruklinting. Or are they with the gods?

Illustration:

Baruklinting lies in wait for unsuspecting prey. Here, this dragon lies camouflaged in the rocky outcroppings.

SOURCES:

Anjas, A. (1985.) The Baruklinting Dragon. Kawanku Magazine (20). Retrieved from http://www.oocities.org/vienna/5385/nagae.html.

Sihombing, Ronny Sahputra, Palupi, Victoria Usadya, and Ragawanti, Debora Tri. (2013). Identification Of Six Elements Of Narrative Used By Third Grade Of Elementary School Students Of Bethany School In Rewriting The Story Of Rawa Pening. Satya Wacana Christian University Institutional Repository. Retrieved from http://repository.uksw.edu/handle/123456789/3402.

The Baruklinting Dragon. (2016.) Indonesian Folklore. Retrieved from http://indonesianfolklore.blogspot.com]/2007/10/ baruklinting-dragon-folklore-from.html.

Souvan and Soutto, the Mekong River Dragons

Background:

Laos is known as a crossroads for trade and migration in southeast Asia. As a result of this history, forty seven ethnicities are recognized in the country. Perhaps two of them originated with this legend. The war described here could be symbolic of warring states that developed in the area, especially conflict over the resources of the Mekong and the Nan Rivers, both of whom originate in Tibet.

Myth:

In a part of China that was once part of the kingdom of Laos, but is now on a Tibetan slope claimed by China, there was the great Kuva Lake, source of the Mekong River. The more common name is Nong Kasae Segnan. Not far away was the source of the Yangtze Kiang, the "Nan" river which was believed to be a playground and haven for dragons, and in it there lived two dragons, Souttoranark and Souvanranark. Each had "ranark" added to his name because both were kings with thousands of followers and admirers. Because the lake was so huge, they did not see each other every day but would visit often as friends, bringing gifts to one another.

On one of these visits, Soutto brought Souvan the gift of elephant meat, which pleased Souvan greatly. Souvan tried, in turn, to find such a huge animal in order to repay Soutto's kindness. However, his hunting trip was unsuccessful, and all he could find was a porcupine. Still, he cut up the porcupine meat into equal portions and sent a share to Soutto. Because the quills were so beautiful, he also sent some of these with the meat. Soutto, however, was offended. He did not understand how small a porcupine is, and thought that the portion he received must be much smaller than everyone else's. When he saw the quills of the porcupine, he was sure that a porcupine was a very large animal indeed. Souvan's messengers tried to explain that very large animals like elephants might have very fine hair, while the meager porcupine's quills are very long, but Soutto turned them away with the gift, convinced that he had been insulted. He also called for a personal meeting with Souvan.

When Souvan heard of Soutto's response, he knew that he had misunderstood the size of a porcupine. Yet he was also hurt that Soutto had not listened to the messengers' explanations, and seemed so quick to think ill of him. Still, he prepared to visit his friend.

At his own palace, Soutto readied himself for Souvan's visit. He insisted to all of his advisors that he had been wronged, and even though many of them saw the king's mistake, none were brave enough to tell him his error in judgment for fear of being punished. When Souvan arrived, Soutto greeted him cordially, but then began to ask why he had betrayed their friendship. Souvan was patient as he tried to convince Soutto that a porcupine was truly smaller than an elephant even though its hairs were larger. He insisted that he was still, and was always, Soutto's dear friend. This merely enraged Soutto, who would not hear it, and who finally stormed away from Souvan.

Back in his own kingdom merely a few days later, Souvan received a declaration of war from Soutto, who insisted that he admit to his wrongdoing or prepare for war. Souvan then realized with a heavy heart that he must assemble his followers and fight his former friend. Soutto came with his dragons, using all of their magical ability and moving so quickly that they muddied the water. When they met and clashed with Souvan's troops, Lake Kuva and the entire surrounding lands became so doused in mud that the sun could not be seen for seven days. On and on the fighting went, and as the churning waters continued to mar the environment all around the lake, small animals began dying or fleeing the lands. Gods in the sky turned away from the area and all its chaos. It seemed that nothing could stop the fighting, with the two sides of the war so equally matched in strength. Finally, the king of the gods threatened both dragon kings, and the fighting ceased.

To make amends, the dragons were given the task of creating two rivers. Soutto was in charge of building the Mekong River, while Souvan was tasked with the Nan River. The king of gods also convinced Soutto of his mistake, showing him the true size of a porcupine. Soutto and Souvan became friends again, and have been ever since. Yet the Mekong and Nan Rivers remember the hatred that gave them birth, and even today it is said that putting water from each river in the same bottle will cause it to crack or shatter completely!

Illustration:

This scene depicts the Mekong River dragons arguing over the size of a porcupine. Their designs are both loosely based on snakes from the family Acrochordidae, which are aquatic specialists found in freshwater ecosystems in tropical Asia and Australia.

SOURCES:

Kaignavongsa, Xay and Hugh Fincher. (1993.) Legends of the Lao: A Compilation of Legends and other Folklore of the Lao People. USA: Geodata Systems, pp. 23-26.

Bakunawa

Background:

The Philippines is a nation with thousands of islands in the western part of the Pacific Ocean. In Philippine mythology, many myths exist about the bakunawa, whose name has been variously translated as "moon eater," "man eater," "eclipse," and "bent serpent." He is one of many creatures in the world created by Bathala, who will be central to another legend in this chapter.

The myth of the Bakunawa can still be seen in children's books, as shown by the Spanish language book "El dragón y las siete lunas" by Joanne de León and the Japanese language book "Otsukisama o nomikonda doragon : Firipin no minwa" by Joanne de León and Masako Fuse. The moon-eating sea serpent is the inspiration behind the short story "The Bakunawa" by Michael Penncavage, and a Bakunawa festival is also at the center of Peter Solis Nery's romance novel, "Love in the Time of the Bakunawa."

Myth:

Also called Tabashie, Bakonawa, Baconaua, or Bakonaua, the Bakunawa had two sets of wings, whiskers on his face, a red tongue, and an enormous mouth. In most stories he is a serpent, but in some he is more a shark. Yet in spite of all these different names and descriptions, one thing is certain: the bakunawa eats the moon.

In the Philippine view of the history of the world, the earth had seven moons instead of one, which the god Bathala created to light the earth at night. Yet Bakunawa rose up and ate many of these moons, and it is still feared that he might eat the last of the moons. Some stories say that dragons of all kinds lived in the sea, and all were allured by the beauty of the moons and tried to eat them. Others say that Bakunawa ate the moons in anger.

Another version of the Bakunara myth tells of his sister, a turtle, laying her eggs on a Philippine island. Every time she laid her eggs, the water followed her and rose up onto the island a bit more, so the people of the island killed her. In his anger and grief Bakunawa ate the moons, and Bathala refused to punish him for it. However, Bathala told the people to bang pots and pans to scare him away, and he was never seen again.

In yet another version, Bakunawa fell in love with a human woman who loved him in return. The leader of her village discovered their love, and burned the house they had built together to the ground. In this story, the Bakunawa swallows the moons out of revenge. And, Bathala does punish Bakunawa, banishing him to the sea. When the waters rise, Bakunawa is trying to come back to his home and his family because the Bakunawa is still alive today. In some places during an eclipse, pots and pans may still be banged together to scare him away and make him regurgitate the moon.

Illustration:

Bakunawa's hunger is such that he tries to consume the whole moon. Described as having an enormous, gaping mouth, his design was based on the basking shark (*Cetorhinus maximus)*, a pelagic species that filter feeds on plankton with a wide open mouth.

SOURCES:

De León, Joanne. (2003.) El dragón y las siete lunas. Japan: Shinseken.

De León, Joanne, and Fure, Masako. (2003.) Otsukisama o nomikonda doragon : Firipin no minwa. Japan: Shinseken.

Fegan, Brian. (1983.) Some Notes on Alfred McCoy, 'Baylan: Animist Religion and Philippine Peasant Ideology.' Philippine Quarterly of Culture and Society 11(2/3), 212-216.

Nery, Peter Solis. (2012.) Love in the Time of the Bakunawa. United States: CreateSpace Independent Publishing Platform.

Penncavage, Michael. (2010.) The Bakunawa. In Dragon's Lure: Legends of a New Age. Danielle Ackley-McPhail, Jennifer Ross, and Jeffrey Lyman, Eds. Howell, New Jersey: Dark Quest, LLC.

Rafferty, Patrick. (2007, March 9.) Visayan-English dictionary. The United States and its Territories: 1870- 1925: The Age of Imperialism. Retrieved from http://quod.lib.umich.edu/p/philamer/ACK6070.0001.001/18?rgn=full+text;view=image;q1=bakunawa.

Tabada, Mayette Q. (2003, March 2.) Tabada: Why doctors become nurses. Babaylan Files. Retrieved from http://babaylanfiles.blogspot.com/2009/07/conversations-tabada-why-doctors-become.html.

Valiente, Tito Genova. (2015 January 1.) A serpent, this earth and the end of the year. Business Mirror. Retrieved from http://search.proquest.com/docview/1644507809.

Ulilang Kaluluwa

Background:

As previously mentioned, Bathala is a central deity in Philippine mythology. The Philippine nation covers a vast area of islands in the western part of the Pacific Ocean. The earliest human remains are dated to 67,000 years ago. Several theories exist about where the earliest folk migrated from - maybe this legend reflects those different population movements.

The Philippines are close to what is defined as the Oceanic region, and sometimes even included in it. This story bears some resemblance to the Samoan myth "Sina and the Eel," in which a king who had shapeshifted into an eel instructed a woman to plant his head after he died, and from it grew the first coconut tree, whose nuts resemble his face as an eel, as well as some local versions of the Agunua tales.

Myth:

Bathala, also known as Bathalang Maykapal, is the supreme god of the Tagalogs, creator of humans and earth, and, some say, King of the Diwatas in Philippine mythology.

One myth relates that there were once three great gods: Bathala, who wandered the earth, Ulilang Kaluluwa, the great serpent of the clouds, and Galang Kaluluwa, a winged god who loved to travel and could often be found near water. Bathala's name translated to "creator," "god," or, sometimes, "goddess." Galang's name came from the word "galang" for "girder" or "support" and "kaluluwa" from the word for "soul" or "spirit," while Ulilang's name came from the words for "orphan" and kaluluwa, "soul." The three gods did not know of each other, until one day from his clouds, Ulilang Kaluluwa saw the god Bathala.

Ulilang did not like seeing another god in an area he considered his own, so he attacked. The two gods fought for three days and three nights until Bathala prevailed. Ulilang Kaluluwa was dead. So great was the wrath of Bathala that Bathala burned Ulilang's body instead of burying it in the ground.

Many years later Bathala met Galang as Galang was travelling around the earth. Bathala greeted him kindly, invited him in for supper, and the two became fast friends. For many years they were very happy together, though Bathala often told Galang of his unrealized dream: to create life for earth.

In the myth, many years went by, but one day, Galang became very ill. When it became clear that he would die, he told Bathala to bury him in the spot where he had burned Ulilang. Eventually Galang passed, and Bathala followed his instructions. Out of the ashes and the burial, grew a coconut tree. The nut of the tree was like Galang's head — round, with two eyes and a flat nose. Even its leaves were like Galang's wings. The trunk, however, reminded him of Ulilang's tough, scaled body. When Bathala looked at the tree he knew it was time for him to create creatures to live on earth. He then went about making all of the creatures of the earth, including the first man and woman. The coconut tree was their source for many of their needs. Their clothing, shelter, and food came from the nuts, leaves and fibers of the coconut tree, considered to be the tree of life.

Illustration:

Ulilang is a creature of the clouds. Real snakes have yet to colonize the skies, but many have conquered forest canopies such as the dog-toothed cat snake (*Boiga cynodon*). Native to Southeast Asia and with skin like tree bark, he was a perfect fit for this dragon. The horns and ears come from the native sambar deer (*Rusa unicolor*).

SOURCES:

Bathala Myths. Read Legends and Myths. Retrieved from www.read-legends-and-myths.com/bathala-myths.html.

Coconut Tree. Read Legends and Myths. Retrieved from www.read-legends-and-myths.com/coconut-tree.html.

Iya, Palmo R. (2016.) EL RENASCIMENTO: UNVEILING THE METAPHORICAL MEANING OF BATHALA. Philippine Association for the Sociology of Religion Journal 2, 1, 1-24.

Jocano, F. Landa. (1969.) Outline of Philippine Mythology. Manila: Centro Escolar University Research and Development Center.

Krupa, Viktor. (2002.) A Mythological Metamorphosis: Snake or Eel? Asian and African Studies 11(1), 9-14.

San Buenaventura, Mariejoy. (2018.) Book Review: Mythological Woman and the Prose Poem in Barbara Jane Reyes's Diwata. Ramkhambaeng University Journal Humanities Edition 37, 1, 193-210.

The Story of Bathala. (2010, November 29.) Bathala. Retrieved from http://bathalanglangitatlupa.blogspot.com/2010/11/ story-of-bathala.html.

The Story of Bathala (Luzon Creation Myth). Tingin-tingin din! Retrieved from http://tingintingindin.weebly.com/ uploads/1/8/3/1/18312609/pre-spanish_creation_stories.pdf.

Tofighian, Nadi. (2008.) José Nepomuceno and the creation of a Filipino national consciousness. Film History 20(1), 77-94.

Trip Mo, Trip Ko. (2012.) University of California Los Angeles Light and Dark.

Wolfgramm, Emil. (1993.) Comments on a Traditional Tongan Story Poem. Manoa 5(1), 171-175.

Biwar's Dragon

Background:

The people of Western New Guinea are thought to be descendants of Austronesian seafarers. Some say that humans came to the area at least 40,000 years ago. The land is mountainous yet tropical with many, many lakes, rivers and islands. Maybe that explains how Biwar and his mother remained isolated while Biwar achieved adulthood. The area of Mimika is part of the Kamoro cultural system which has a matrilineal emphasis.

In some versions of this story, the hero's name is Mirokoteyao, and he grows with extraordinary speed, learning to hunt prolifically in just a week. When he decides to attack the dragon, he creates a fake village scene and starts a fire to lure the dragon. Once he defeats him, he throws part of his body to the interior, where it became the highland people. Another part became New Guinea. Still other thrown parts became Europe and Asia. This story is still performed by Kamoro people, and some now read into it a political message - that the dragon represents multiple forms of oppression, and the Kamoro people will overcome them.

Myth:

Today, this legend is still passed on about the village of Mimika and a dragon. The people of Mimika were very industrious people, always working hard together to sustain their community. One day, they organized a journey to find sago, sometimes referred to as saksak, rabia or saga. Sago is the spongy center of a tropical palm tree. It is used to make flour and cooked in many forms to feed the people.

Twelve boats were prepared to go in search of sago. The journey lasted three days and much sago was gathered and filled the boats. As the boats headed home, great waves crashed against the sides. The waves were being caused by a dragon swinging its tail! Many of the village people drowned that day, but one pregnant woman clung to a tree. She ate the fruit and roots of the tree to survive until she gave birth to her son, Biwar. With his father and so many villagers gone, she decided to raise Biwar on her own.

As Biwar grew to a young man, he became skilled in many areas such as making fire and weapons, and especially trapping. He also sang and played the tifa, a drum of wood, carved on the sides with holes in the middle and a skin, usually deer, covering the top. Biwar became a great provider. One day, he came home with many fish. His mother asked him where he had caught them. When he said it was upriver, she sadly told him the tale of the loss of his father and the other villagers. She did not want him to return to that part of the river.

But Biwar believed he could slay the dragon. So he set a trap in the mouth of a cave. Then he used his skills at drumming and singing to lure the dragon to the cave. When the dragon placed its head inside the mouth of the cave, Biwar released his traps. Ropes released many weapons, arrows, spears, clubs, even daggers. They all hit the head of the dragon so hard that the dragon died. Biwar went home and told his mother what he had accomplished.

The very next day, Biwar and his mother sailed home to Mimika and were warmly welcomed by the village. They were celebrated and given ceremony to acknowledge their place in the village and their accomplishment in keeping the village safe from the dragon in the river.

Illustration:

Biwar's dragon is an aquatic dragon with a tail strong enough to topple boats. While not nearly as powerful, the asian water monitor (*Varanus salvator*) is a lizard native to southeast asia who is well adapted for swimming. Reaching up to 7 feet long, their sinusoidal locomotion is reminiscent of serpents.

SOURCES:

Biwar. Indonesian Folklore. Retrieved from http://indonesianfolklore.blogspot.com/2008/02/biwar.html.

Coconut Tree. Read Legends and Myths. Retrieved from www.read-legends-and-myths.com/coconut-tree.html.

Harple, T.S. (2001). Controlling the Dragon: An ethno-historical analysis of social engagement among the Kamoro of South-West New Guinea (Indonesian Papua/ Irian Jaya). Unpublished PhD thesis, Australian National University, Canberra.

Rustan, Mario. Biwar Kills a Dragon. Indonesian Myth. Retrieved from http://www.st.rim.or.jp/~cycle/MYdragonE.HTML.

Lac Long Quan

Background:

Legends still exist of the origins of the Vietnamese people. Some scholars see this story as emerging during the transition from a matrilineal system to a patrilineal system. However, others have argued that the two coexisted for many years and in some places still do to this day. While Chinese Confucianism pushed a patriarchal system, it affected mostly upper class Vietnamese women, while in the Le societies of the 15th to 18th centuries, working class women retained all property within marriage, could divorce, and by law, inherited from their parents (and to their offspring). They could also assert rights to their children.

Even in the 18th century, missionaries reported that women in Vietnam could claim half of their children in a marital split, and in some villages today, it is still understood that upon a separation, the mother has rights to the firstborn son and the father has rights on the second-born son. Some scholars contend that the equal division of the 100 sons within the myth signifies an early example of this bilateral kinship system, as this tale is at least as old as the 14th century.

Within the myth, it is also notable that Lac Long Quan gets his royal status from his father but his draconic nature from his mother, again signifying ties to both his paternal and maternal lineages. His own name, Lac Long Quan, follows Chinese naming conventions (which favored patrilineality due to Confucian influence), even using the Chinese word for dragon ("long"), but Au Co's name follows traditional Vietnamese naming conventions.

One version begins with King Minh, thought to be a descendant of Than Nong (Shennong in Chinese). Shennong is believed to be the son of a princess and a heavenly dragon. Shennong is also known as the "Divine Farmer." Many legends exist in Asia of ancestors who are dragons, and Vietnam is no exception.

Myth:

King Minh met Lady Vu, an immortal, as King Minh was travelling through the south. They had a son, Loc Tuc, whom the King wanted to inherit the throne. But Loc Tuc wanted his brother (some say half brother) to inherit it, so King Minh split the kingdom in two with King Nghi inheriting the North and Loc Tuc inheriting the south.

And that's where Loc Tuc, who some call King Kinh Duong Dragon, meets Long Nu, a daughter of another dragon king. They had a son, Sung Lam, who later became known as the Dragon Lord of the Lac or Lac Long Quan, the grandson to King Minh.

Lac Long Quan could live underwater, but more importantly, he was so good at ruling the south - introducing agriculture and a social system that brought sustainable order - that sometimes he would have to travel to make sure that everything would still work well. Lac Long Quan defeated the fish demon, a nine tailed fox and an evil giant tree spirit to protect his people.

Lac's cousin, King Lai, the son of King Nghi of the North was married to Au Co, a princess, some say an immortal, from the high mountains. One day,King Lai decided to go travelling, leaving Au Co and forgetting to return.

It was at this time, that the North (Chinese territory today) was oppressing the South (Vietnam) and it was also one of those times that Lac Long Quan was in his underwater realm, but Lac's people knew that they had but to call for him and he would return. When the people of the South called to Lac, because of King Lai's oppression, Lac immediately appeared.

And what he saw was Au Co. Her beauty stunned him and he fell in love. Au Co returned the love and followed Lac to one of his palaces willingly.

When King Lai finally reappeared, Au Co was nowhere to be seen and try as he might, he could not find her. Lac kept creating magical obstacles to bar Lai's way and Lai finally gave up.

Lac and Au Co gave birth to a sack of 100 eggs. Each egg grew into a handsome son. But Lac still needed to travel back and forth to his underwater realm. In the myth it is said that after much time, the two decided to split and each took 50 sons to their home: Lac to the underwater kingdom and Au Co to the high mountains.

This is the myth of the 100 Vietnamese clans and their dragon origins.

Illustration:

Great ruler of the south, Lac Long Quan is drawn here using references from classical southeast asian art. The long sinuous body of this dragon is inspired by the thin and regal asian vine snake, *Ahaetulla nasuta*.

SOURCES:

Le, C. N. Vietnam: Early History and Legend. Asian-Nation. The Landscape of Asian America. Retrieved from http://www. asian-nation.org/vietnam-history.shtml.

Nguyen, Dieu Thi. (2013.) A mythographical journey to modernity: The textual and symbolic transformations of the Hùng Kings founding myths. Journal of Southeast Asian Studies 44(2), 315-337.

The Legendary Origins of the Viet People. Vietnam Culture: Brings Vietnamese Culture to the Rest of the World. Retrieved from http://www.vietnam-culture.com/articles-47-4/The-Legendary-origins-of-the-Viet-people.aspx/.

The Origin of Viet People. (2013, April 21.) Vietnamese Myths and Legends. Retrieved from https://sites.google.com/site/ vietnamesemythsandlegends/the-origin-of-viet-people.

Van Ky, Nguyên. (2002.) Rethinking the Status of Vietnamese Women in Folklore and Oral History. In ViêtNam exposé : French scholarship on twentieth-century Vietnamese Society. Gisele Bousquet and PierreBrocheux, Eds. Ann Arbor, Michigan: The University of Michigan Press.

Yu, Insun. (1999.) Bilateral social pattern and the status of women in traditional Vietnam. South East Asia Research 7(2), 215-231.

Two Dragons Who Stole the Sun and the Moon

Background:

Taiwan is a complex geographic island with several places where, over time, tectonic plates have collided. Earliest human remains are approximately 20,000 -30,000 years old. Volcanic eruptions are known throughout the island's history. There were likely indigenous groups there before the Austronesian people arrived. It is not unusual for humans to remember natural disasters through myth and legend.

Myth:

In Taiwan, the legend is still passed down from the Indigenous peoples and cultures of the Shao/ Tsou about how they were living on the island in the middle of Sun Moon Lake, with agriculture, hunting, gathering and fishing, when a white deer led them to the side of the lake where there was much more farmland. They lived on the side of the lake for many years, growing millet, upland rice, taro, sweet potato and other crops such as ginger.

But one fine day, as the people were working in their fields, a very large BOOM was heard. The land shook and the sun disappeared. The people managed to find their way home and that night they decided to work by the light of the moon when BOOM, the land shook again and this time the moon disappeared. The world had become a frightening, very dark place.

Each day and night, there was no light. Without light, the crops withered and died. The fish, who had been so plentiful, hid in the bottom of the lake and animals seemed to also wither away. Without light, nothing grew. The people did not know what to do.

However, a woman, ShuiShwJie and her husband, DaJianGe, thought about their plight and realized that they needed to come up with a plan. They came to the conclusion that the sun and moon were lost somewhere, maybe in a deep valley, so they determined to go in search of them. They went over rivers and through many wooded areas. They climbed mountains, but could not find any sign of light, until, finally, ShuiSheJie saw a distant light on top of a lake. They ran towards the lake. But when they reached the lake, what did they find? Two very large dragons were playing with the sun and the moon, their own personal fireballs. ShuiSheJie and DaJianGe were angered by this. Stealing the light from the people, the plants, the animals and the fish, just to play ball. They wanted to steal the sun and moon back, but were afraid of the two large dragons. The dragons were very large and very fierce.

So the couple sat and thought and thought. At one point, they noticed that white smoke was coming up from under the rocks that they were sitting on, so DaJianGe pushed the rock aside. Under the rock was a very long, very deep and very narrow underground tunnel. The smoke was winding up along the tunnel. As they climbed down, it got darker and damper, but finally they came upon an old woman tending a cooking fire.

They greeted the old woman who was very surprised to see them. She said that she had not seen other humans in a very long time.

"Many years ago when I was working in the field, the dragons captured me and brought me here. They would not let me leave this place. They also forced me to cook them meals."

(archiver.rootsweb.ancestry.com/th/read/FOLKLORE/2000-12/0976781019)

ShuiShelJie and DaJianGe told the old lady about how the dragons had stolen the sun and the moon and the desperate situation aboveground. But the old lady said that they would not be able to defeat the two dragons. The only help that the old lady had was a tale that the two dragons were afraid of a pair of golden scissors and a golden ax under the Ali Mountain.

So the couple went off in search of the golden scissors and the golden ax. Since the ax and scissors were supposed to be under the mountain, the two began to dig with the strongest wooden sticks that they could find. They dug and dug until they saw gold shining. It was the scissors and the ax. They took the golden tools and went back to the lake where the dragons were still playing ball with the sun and the moon.

ShuisheJie threw the golden scissors and they went straight at the first dragon, cutting its body into many pieces. The DaJianGe threw the ax and it cut off the head of the second dragon. Now however, the sun and the moon were floating on top of a very bloody, red lake. They saved the old lady who had one more piece of advice: in order to get the sun and the moon back into the sky, the couple would need to eat the dragons' eyeballs. This would make the couple grow very, very tall. Then they could throw the sun and the moon high into the sky. They swam into the lake, found the eyeballs, ate them, grew tall, but weren't tall enough. The old lady saw the problem and suggested that the couple use two tall palm trees to hold the sun and the moon up in the sky.

The couple were now of giant stature. They took one palm tree, slowly lifted the sun up and into the sky. Then they did the same with the moon. It worked. Light shone down upon the land, plants and flowers bloomed. Animals and fish came to life again.

Just one problem remained. The couple were worried that the dragons might come to life again and cause more problems, so they stayed and guarded the lake. Slowly they became the mountains around the lake, now called, DaJian Shan and ShuiShe Shan.

The people created the "Holding Ball Dance" to thank the couple, which they dance to this day.

Illustration:

Another pair of aquatic dragons, these are drawn with the body shape of an eel. Here they are playing catch with the sun and the moon, with two large mountains in the background. Also of note, Taiwan is in a seismically active zone on the Pacific Rim of Fire. The most recent eruption of a volcano on Taiwan was the Datun Mountain, 5,000-20,000 years ago, though the Datun volcano group is still active. Its first historically recorded earthquake was in 1624.

SOURCES:

Taiwan volcano: active, dormant, or what? (2009, November.) The Volcanism Blog. Retrieved from https://volcanism.wordpress.com/2009/11/02/taiwan-volcano-active-dormant-or-what/.

Thao. Digital Museum of Taiwan Indigenous People. Retrieved from http://www.dmtip.gov.tw/web/en/page/ detail?nid=13.

The Legend of Sun-Moon Lake. (1996, September 7.) Folk Stories of Taiwan. Retrieved from http://www.taiwandc.org/folk-sun.htm.

Shung Ye Museum of Formosan Aborigines Guidebook. (2017.)

Princess Manora

Background:

Thailand is a nation that shares a peninsula with Laos, Vietnam, Cambodia and Malaysia. Like much of the region it is a complicated system of rivers, mountains and river valleys. As with most of Southeast Asia, human habitation can be dated to at least 40,000 years ago.

While the feminine bird-creatures called kinnaree, or kinnari, are often equated to ideal wives in Buddhist literature, they have many more legendary characteristics. For example, they are thought to be able to guide lost people through woods, and are associated with spiritual power. They are still depicted in modern artwork; contemporary Thai artist Chakrabhand Posayakrit completed a mural at the Wat Tri Thotsathep Worawiham temple in Bangkok entitled "Kinnari and Kinnara."

Like other countries in the area, the cultural system was strongly influenced by Hinduism and its patriarchal kinship system, and several areas have recently been influenced by Islam. However, female-oriented spiritual practices based on this very myth have endured, despite official attempts to ban them. A type of trance theater practiced in Thailand and Malaysia was begun by two women, and since only women originally told this story, some see it as having connections to women's spiritual power and women's rights. This seems particularly poignant given that Manora uses her wits to gain her freedom in one version of the story. Today, men are involved in the telling of this myth, but some dress in female attire while performing to enhance their spiritual power.

The practice of abducting kinnaree in the myth, and the very real practice of abducting women associated with the spiritual trance of Manora, then, are far more than the act of collecting attractive women. These women would have enhanced the perceived power of the rulers they served under, and were thus political acquisitions. A king could become godly through his association with these powerful women.

This story may have traveled, with Japanese, Swedish, and North American Indigenous tales bearing strong similarities to the myth. The Japanese Noh play Hagomoro, for instance, bears a resemblance to this tale.

Myth:

A great love story is still told in Thailand. It is the legend of Princess Manora and Prince Suthon and, of course, involves a dragon.

In the mythical kingdom of Mount Grairat live a king and queen with seven beautiful daughters known as the Kinnaree. These were young women, some say goddesses, who were known for their excellence in singing and dancing and the fact that they could sprout wings and fly when they wished. They often flew to their favorite, secret bathing pool.

In the nearby kingdom of Panchala Nakhon, the king and queen had only one child, a son, Prince Suthon. Suthon's parents wanted to find the perfect wife for their only son. After all, Suthon was handsome and the most amazing archer, so good that he was referred to as the Good Archer. His parents interviewed many young women, but none fit their requirement, "beautiful as the rose and as gentle as a doe". (gotoknow.org)

It happened that the best hunter in Panchala Nakhon, Pran Boon (Prahnbun) discovered the secret bathing pool of the Kinnaree and seeing their great beauty, and their skills of singing, dancing and flying, decided that he must capture one for Prince Suthon. Some say Pran Boon sought the advice of the great dragon and convinced that dragon, a Naga, to lend him a magical rope which could capture the Princess. Some say Pran Boon "borrowed" the rope. Whatever the truth, Pran Boon was able to return to the bathing area and capture Princess Manora and bring her to Prince Suthon and her parents.

Pran Boon was rewarded, but that is not the end of the myth.

Prince Suthon immediately fell in love with Princess Manora, but her parents were not thrilled with how events had transpired. Prince Suthon spent several years accomplishing many feats ordered by her parents in order to prove his love and devotion to Princess Manora. When her parents were finally satisfied with Suthon's acts of devotion, Suthon and Manora were married and, some say, lived happily ever after.

Other versions of the story say that the hunter captured Manora with a rope, after first stealing her feather robe which allows her to fly, and gave her to the prince. Though they lived happily and had children, other nobles resented her and threatened to kill her when he left on an expedition. The resourceful Manora pretended that she wanted to perform a dance, and thus was given her feather robe back. She uses it to fly away from the danger, and regain her freedom.

Illustration:

Depicted here is the naga who lends the hunter a magical rope to capture princess Manora. His head and body are loosely based on the king cobra (*Ophiophagus hannah*), a venomous snake native to southeast Asia which can reach lengths of nearly 18 feet.

SOURCES:

Giordano, John T. (2005.) Kinnari: On the Space between Traditional and Corporate Myths. Prajna Vihara 6(1), 90-107.

Manora: A Thai Legend. Manora Property. Retrieved from http://www.manoraproperty.com/136-manora-a-thai-legend.

Plowright, Poh Sim. (1998.) The Art of Manora: an Ancient Tale of Feminine Power Preserved in South-East Asian Theater. New Theatre Quarterly 14(56), 373-394.

Pra Suthon-Manora. Translated by Marion Davies. Gotoknow. Retrieved from https://www.gotoknow.org/posts/458723.

Princess Manora. Angelgenie. Retrieved from www.angelgenie.com/index.php?cPath=20.21.

Cultural Questions:

1. In Lac Long Quan, we again see familial conflict and power struggles, as well as connections to Chinese dragon myths. Consider the implications.
2. Bathala is a main character in two of the myths. Discuss Bathala and his function in the Philippine cultural systems. What is the significance of coconut trees?
3. What concepts or emotions does the Bakunawa represent?
4. Could the Two Dragons Who Steal the Sun and the Moon represent ecological events or historical forces? How do these two dragons differ from other Asian dragons?
5. In the Mekong River Dragons, we again see the possibilities of ecological and/or historical events. Discuss the options. Could this reflect two warring populations?
6. Discuss the familial conflicts in the Princess Manora story. Was it ever really a love story if the princess was kidnapped?
7. How might the different versions of the Princess Manora story be used in connection to women's rights and their place as spiritual icons?
8. As the chapters move further east, discuss the humanization of the dragons.

CHAPTER 4: CHINA

Background:

Evidence of the pre-human form known as Homo Erectus dates back to 1.7 million years ago in China. This was Yuanmou Man. Another Homo Erectus form, Peking Man dates to 500,000 years ago. Clearly China was occupied very early in human history.

The earliest domesticated crop appears to be millet dating back to 9000 years ago. From then on, Chinese stories tell of the sun, the moon, the stars and all beings which might affect agriculture. Water, droughts or floods figure prominently. Chinese culture claims to have 3500 years of written history. At some point, one of the most common deities became the dragon, the long or lung.

During the Paleolithic, a matriarchal clan system managed a primitive farming cultural system with craft specialization. During the Neolithic, the cultural system changed over to one of patriarchy, evident in the many dragon myths.

The greatest leaders either were dragons or children of dragons. One example is Shennong, the second of three great medically oriented emperors. The story says that Shennong was the son of a princess and a heavenly dragon, that he was born around 2800 B.C./B.C.E. or 4800 before present, about the same time as one of the earliest Neolithic carvings of a dragon, the Pig Dragon. Chinese cultural history impacted its neighbors, Japan, Korea, Vietnam, Mongolia and others.

While Chinese dragons are associated frequently with royalty, there also many stories where they interact with common people. For instance, a short story about a scholar who sees a dragon in miniature form and puts on his most respectful outfit to carry it outdoors demonstrates that dragons could show favor to humans who were not in the upper classes. Again, water is a prominent feature of many of the myths as well as cultural/historical beginnings and dragons are seen as the spirits of those waters.

Illustration:

Pig Dragon, Hongshan culture, from approximately 5,000 years ago. Whatever the truth of the origins of Chinese peoples and cultures, dragons remain a constant in the stories.

SOURCES:

Latini, David, Ed. (2018.) Epic Tales: Chinese Myths and Tales. London, UK: Flame Tree Publishing.

Four Rivers of China

Background:

Changes in climate are reflected in changing cultural adaptations. The following dragon legends could well reflect climatological changes such as flooding, droughts and other natural events. In fact, we know there was an era of flooding and droughts in northeast China, home to many Neolithic cultural systems, around 6000 before present. At 4000 before present, another abrupt drought began in the eastern margin of the Tibetan Plateau, the western part of the Chinese Loess Plateau and in the vast Inner Mongolian Plateau.

Myth:

Long ago, and still today, a story is told of the origins of the four major rivers of China; The Yangtze, the Yellow, the Pearl, and the Heilonggiang or Amur River. The land was drying out. It was a time of great drought. There was very little rain. There were no rivers and lakes for the people to rely on. They desperately needed rain for their crops.

There was an Eastern Sea where four large and beautiful dragons lived: the Long Dragon, the Yellow Dragon, the Black Dragon, and the Pearl Dragon. These dragons swam in the sea and flew into the air. They controlled the secondary or "sweet" waters, and loved to play among the clouds.

One day, the dragons flew into the air and noticed that the land and the people on it were suffering from a severe drought. The dragons felt very bad that the people were suffering so. There were many women too thin to live, trying to feed babies who were dying . The people were eating anything they could, dry grass, dried up crops, even dry dusty clay. It was a disaster.

The four dragons decided to fly to the home of the heavenly Jade Emperor, the very powerful deity who was supposed to be taking care of everything on earth, in heaven, all of the people, even the sea. When the dragons reached the Jade Emperor, they saw that the Emperor was busy watching fairies dance and frolic. He was annoyed that the dragons were disturbing his fun. When the dragons begged the Emperor to bring rain to the people, he said that he would, but as soon as the dragons were gone, the Emperor forgot about the suffering of the people and the rain.

Many more days went by without rain. The dragons became very distraught. The land, the plants, the people were dying and still no rain came from the Jade Emperor.

The dragons came up with a plan. There was plenty of water in the sea, and even if the Jade Emperor found out and was displeased, the dragons just couldn't watch the people suffer any more. So, they decided to take water from the sea and spray it into the air where it would fill the clouds up and then it would rain. The plan went amazingly well. The dragons flew back and forth from the sea to the clouds over the land and soon it rained. The people were saved. The crops grew again and life was good.

However, there was a sea god who was not at all happy that the four dragons had taken water from his sea without asking. He complained to the Jade Emperor. The Jade Emperor was not at all happy that the four dragons had done this without his permission. In fact, the Emperor was very, very angry.

The Jade Emperor saw that it was the four dragons who had made him so mad. He sent a group of heavenly generals to bring the four dragons to him for punishment. To make sure that the dragons would never escape, the Jade Emperor had the Mountain God take four mountains, and place one on each dragon. Now they could never escape back to the sea. But the dragons have never regretted saving the people. In fact, as the story goes, the four dragons turned themselves into four rivers that ran through the mountains down to the sea to make sure that the people would always have water. And that is how the four rivers were formed.

Illustration:

The Dragons of the Four Rivers save China by turning into rivers. Each dragon is based on four sea snake species found in the Pacific Ocean: the olive sea snake (*Aipysurus laevis*), the banded sea snake (*Laticauda colubrina*), the yellowbelly sea snake (*Hydrophis platurus*), and the horned sea snake (*Acalyptophis peronei*).

SOURCES:

A Chinese Tale. (2014.) Cultural China. Retrieved from traditions.cultural-china.com/en/211T11672T14506.html.

Fredericks, Deby. (2012, March 22.) The Four Dragons, a Chinese folk story. Wyrmflight. Retrieved from https://wyrmflight.wordpress.com/2012/03/22/the-four-dragons-a-chinese-folk-story/.

Ke, Yuan. (1991.) Dragons and Dynasties: An Introduction to Chinese Mythology. Translated by Kim Echim and Nie Zhixiong, London: Penguin.

Latini, David, Ed. (2018.) Epic Tales: Chinese Myths and Tales. London, UK: Flame Tree Publishing.

Liu, Fenggui and Feng, Zhaodong. (2012.) A dramatic climatic transition at ~4000 cal. yr BP and its cultural responses in Chinese cultural domains. The Holocene 22(10), 1181–1197.

Sanders, Tao Tao Liu. (1983.) Dragons, Gods & Spirits from Chinese Mythology. New York: Schocken Books.

Werner, E. T. C. (1922, 1994.) Myths and Legends of China. Dover Publications.

Gong Gong

Background:

Many legends are still told of Gong Gong, the God of Water often described as an immense black dragon, some say he had the tail of a water serpent. Since black is the color associated with water in traditional Chinese culture, this description is a fit one. Some myths also describe him as having red wavy hair, in keeping with his passionate disposition. He smashed his head against the mythical Buzhou Mountain, forever tilting the lands so that all of the rivers flow in the same direction. All of the stories relate to his abilities to control water.

Myth:

Gong Gong had tremendous power to control water. But he was just a minion in the heavenly court of the Jade Emperor. This did not sit well with Gong Gong, so he staged a rebellion and unleashed vast amounts of water upon the earth causing massive amounts of damage to all and sundry. However, he was opposed by the god of fire, Zhu Rong and an intense battle ensued which involved all of heaven and earth. In the end, Zhu Rong was victorious and Gong Gong threw himself at the closest mountain, one of the pillars that was holding up the earth. This didn't hurt Gong Gong much, but it did unleash even more destruction on the earth. It created a hole in the mountain from which erupted fire and more flood and knocked the earth off of its axis. This could have been the end of all living things on earth.

However, Nuwa, known as the goddess of creation, came forth to save the children of earth, her children. Some say she used the body of a turtle to keep the earth on its axis. Many stories are told of how she took special pebbles from a river and magically fixed the heavens. All of her repairs did not completely undo the destruction caused by Gong Gong. That's why the rivers all flow to the east and the sun and the moon flow from east to west. For his own part, Gong Gong lost much of his status and has hated fire since his fight with Zhu Rong.

There is another myth about Gong Gong where he is not the destroyer but the dragon god who teaches people flood control. He figured out how to make embankments which could direct the flow of water and this greatly improved agricultural production. But Gong Gong's people were overcome by Yu the Great. After that, we find stories told of Gong Gong the destroyer whose release of floods upon the earth is countered by Yu, a golden dragon and son of a white horse named Gun. Yu is then credited with teaching people methods of flood control. And it is Yu who is credited with making a turtle to carry the earth and organizing other spirits, such as faeries, along with the dragon, Ying Long, to create paths for the flood waters to flow to the sea.

History and mythology are (re)written by the conquerors.

Illustration:

The dark Gong Gong is drawn here with the skin of *Xenodermus javanicus*, a dark gray snake found in southeast Asia. This snake is nocturnal and is often called the 'dragon snake' due to the prominent rows of keeled scales down its back.

SOURCES:

Gong Gong (the God of Water). (2014.) Cultural China. Retrieved from http://traditions.cultural-china.com/en/13Traditions1394.html.

Gong Gong, Nuwa, and the Fragile Nature of Life. (2011, May 19.) ferrebeekeeper. Retrieved from https://ferrebeekeeper.wordpress.com/2011/05/19/gong-gong-nuwa-and-the-fragile-nature-of-life/.

Ke, Yuan. (1991.) Dragons and Dynasties: An Introduction to Chinese Mythology. Translated by Kim Echim and Nie Zhixiong, London: Penguin.

Latini, David, Ed. (2018.) Epic Tales: Chinese Myths and Tales. London, UK: Flame Tree Publishing.

McLeish, Kenneth. (1996). Myth: Myths and Legends of the World Explored. New York, NY: Faces On File, Inc.

Sanders, Tao Tao Liu. (1983.) Dragons, Gods & Spirits from Chinese Mythology. New York: Schocken Books.

Zhao, Qiguang. (1989.) Chinese Mythology in the Context of Hydraulic Society. Asian Folklore Studies 48(2), 231-246.

Nian

Background:

Nian means "year" in Mandarin, and this dragon is closely associated with celebrations of the new year. In ancient Chinese hieroglyphic portrayals, Nian was a man carrying crops on his back, perhaps a sign of a spring harvest. The Chinese Lunar New Year is alternately described as a Spring Festival. In this older version, Nian was a deity who killed a monster. In later renditions of the tale, Nian himself is the monster and harms people unless they scare him away.

The tale of the dragon Nian is retold in several children's books, including "The Nian Monster" by Andrea Wang, "Nian, the lunar dragon: a famous Chinese legend retold by Sofia Goodsoul and Marina Kite," and the upcoming book "Nian, the Chinese New Year Dragon by Virginia Loh-Hagan. Some scholars have recently advocated for a more ecologically friendly approach to the practices that originated as a way to scare the dragon, as the fireworks.

Some say this story focuses specifically on Peach Blossom Village in Shanghai

Myth:

Many, many legends are told of Nian, a fierce creature with the head of a lion, razor sharp teeth, a mean growl and the body of a bull. He lived in the mountains above a village usually eating whatever came near. At the end of winter, when food was scarce, however, he would come down off of the mountain and raid the village for food, sometimes even eating people. Apparently small children were especially tasty. The villagers were justifiably terrified of him, but an elder taught the villagers that Nian was greatly afraid of fire, the color red and loud noises.

So, one spring the villagers decided to hang red lanterns and burned bamboo which made a crackling noise to scare Nian away. It worked very well. So every year, at the beginning of the lunar calendar, the traditional time of the beginning of the new growing season, spring, villagers throughout China set off fireworks now instead of burning bamboo and put red lanterns on their door and generally make sure to scare off Nian who is said to still be hiding in the mountains. Red became the symbol of good luck, wealth and prosperity.

This is probably where the custom of the dragon dance originated. Several men will hold up the paper body of a dragon, usually red, with lanterns at either end, and dance through the streets making a great deal of noise. Some say it is to rouse the earth to life after winter, but some say it also keeps Nian away.

Illustration:

The fierce Nian is a mixture of lion and bull parts. This illustration uses stylized elements from Chinese sculpture to render the mane and horns.

SOURCES:

Goodsoul, Sofia. (2015.) Nian, the lunar dragon: a famous Chinese legend retold by Sofia Goodsoul andMarina Kite. Clayton South, Vic. Volya Press.

Holloway, April. (2014, January 31.) The Origin of Lunar New Year and the Legend of Nian. Ancient Origins: Reconstructing the Story of Humanity's Past. Retrieved from ancient-origins.net/myths-legends-asia/origin-lunar-new- year-and-legend-nian-001289.

Loh-Hagan, Virginia. (expected 2019.) Nian, the Chinese New Year Dragon. Ann Arbor, Michigan: Sleeping Bear Press.

Qingdao China Guide. (2016.) Retrieved from qingdaochinaguide.com.

Sanders, Tao Tao Liu. (1983.) Dragons, Gods & Spirits from Chinese Mythology. New York: Schocken Books.

Wang, Andrea. (2016.) The Nian Monster. Kirkus Media LLC.

What Is the Legend of Nian? (2015.) Quora. Retrieved from https://www.quora.com/What-is-the-legend-of-Nian.

Ye, Chao, Chen, Ruishan, and Chen, Mingxing. (2016.) The impacts of Chinese Nian culture on air pollution. Journal of Cleaner Production 112, 1740-1745.

Zhishu, F. (2017.) A Chinese renaissance. Nature Plants 3(1).

Pai Lung

Background:

The temple of the white dragon, Pai Lung, is located in Mih Yun Hsien, a city among the mountains north of Beijing. The myth presents cultural values concerning gender in the Chinese cultural system. There may also be some significance to Pai Lung's bright white color, as white was the traditional mourning color for the Chinese, and it is unique as a color for a dragon within Chinese mythology.

Myth:

Long ago, on a stormy, stormy night, an old man knocked at the door. A young woman answered and allowed the old man to come in from the storm, into her family home. The next morning, the storm was over and the old man went on his way. Then the girl and her parents discovered that the girl was pregnant. Was this a supernatural act? An assault of some kind? In either case, her parents expelled her from the family home. Some say the girl even had to leave her hometown. She had to care for herself as the child inside her grew.

Finally, she gave birth, but it was just a ball of flesh. Whether in sorrow or in disgust, we may never know, but the girl threw the lump of flesh into the ocean whereupon the ball of flesh turned into a magnificent white dragon with five toes on each foot. The dragon soared into the air causing a huge hurricane. The great storm and Pai Lung's transformation were dramatic events. His mother fainted dead away, and never woke up again, but her reputation changed from an outcast to the revered mother of the white dragon and she was buried with great honor. To this day, her tomb attracts the faithful.

Illustration:

Pai Lung looks over his new form as a white dragon. The rare Chinese crocodile (*Alligator sinensis*) gives its likeness to this dragon. The Chinese crocodile is usually dark in color, but it is possible for them to be born with albinism.

SOURCES:

Bane, Theresa. (2015.) Encyclopedia of Beasts and Monsters in Myth, Legend, and Folklore. Jefferson, North Carolina, USA: McFarland & Company, Inc.

Dekirk, Ash. (2006.) Dragonlore: From the Archives of the Grey School of Wizardry. New Jersey: Career Press, Inc.

Pai Lung. The Sovereign Lair. Retrieved from www.angelfire.com/dragon2/thoth/pailung.html.

Pai Lung. (2015.) Mythology Dictionary. Retrieved from www.mythologydictionary.com/pai-lung-mythology.html.

Rose, Carol. (2000.) Giants, Monsters, and Dragons: An Encyclopedia of Folklore, Legend, and Myth. New York, USA: W. W. Norton & Company, Inc.

Tchang's Pearl

Background:

Tchang and his mother lived on the edge of a lake, further emphasizing the connections between dragons and bodies of water in Chinese culture. The numbers in the story are important, as three is a fortunate number. Forty nine, because it is seven times seven - and seven is associated with ghosts and death - is important for traditional Chinese funerary practices.

Myth:

Long, long ago, a myth told about a boy and his mother who were very, very poor. No matter how hard they worked, they could not make enough to live on. In frustration with the situation, the boy, Tchang, decided to embark on a journey to go to the greatest god and ask why he and his mother were always in such dire straits.

Tchang walked and walked. The story is told that he walked for 49 days. He was starving. He finally came to a farmhouse where a woman and her mute daughter lived. They took him in and fed him. When they learned where he was going, the old woman asked if Tchang would ask the greatest god why her daughter could not speak. Tchang agreed to ask that question.

Tchang walked on and on. Some say he walked for another 49 days and was again near starvation when he came to an old man and his barren orchard. The old man took him in and fed him and when he learned about Tchang's mission, he asked Tchang to ask the greatest god why his trees no longer bore fruit. Tchang agreed.

Tchang walked on and on again. Some say it was yet again another 49 days (49 days is the number of days in the funeral ceremony of many in China; representing death and rebirth). Then he came to a river which was wild and raging. Tchang thought his journey was done, but, suddenly, a dragon with very small wings (too small to let the dragon fly) rose from the wild waters. The dragon had a pearl on his forehead. He asked Tchang if he wanted to cross the river and why. Tchang explained his quest. The dragon thought this justified taking Tchang across the river, but asked Tchang to ask the greatest god why he, the dragon, didn't have adequate wings. Again, Tchang agreed to ask the question.

Finally, Tchang reached the place of the greatest god. Some say he looked like an old man with a long flowing beard. Hearing about Tchang's long, long journey, Tchang was told that he could ask 3 questions. So he asked the three questions of the three who had aided him, but not his own. When he travelled back, first, he came to the dragon. The answer to the dragon's question was that the dragon needed to do a good deed. The dragon gave Tchang his pearl as a thank you. Then he took Tchang across the river. This was a good deed. The dragon's wings grew and he was able to fly.

Then Tchang came to the old man with the barren orchard. Tchang said the answer to the old man's question was to dig under one of the trees and there were several jars of the purest water. The old man used one to water his trees and they quickly blossomed and bore fruit! The old man gave Tchang one of the jars of water in thanks.

Then Tchang came to the house of the woman and her mute daughter except that now she could speak! Tchang took the young woman as his bride and brought her home, but when he reached home, he found that his mother had gone blind. He gave his mother the dragon's pearl to hold and wonder of wonders, his mother could see again. The dragon's pearl and the amazing jar of pure water made their land fertile again and every year, the story is told, that the dragon would come to visit.

Illustration:

The Asian arowana (*Scleropages formosus*) is famous for its resemblance to a chinese dragon; it is bright red, with wide scales and whiskers. Here, the dragon from Tchang's Pearl is inspired by this fish, augmented with the horns of a domestic cow and very small wings.

SOURCES:

Blackford, Andy. (2014.) Dragon Tales. Oxford: Oxford University Press.

Niles, Doug. (2013.) Dragons: the Myths, Legends, & Lore. Avon, Massachusetts: Adams Media.

Radhakrishnan, Reeja. (2014, June 27.) Tchang and the Dragon's Pearl. The New Indian Express. Retrieved from www.newindianexpress.com.

Zhan, Jade. (2019.) Living By the Numbers. Shen Yun Performing Arts. Retrieved from https://www.shenyun.com/blog/ view/article/e/GVxkb2N-pEY/chinese-lucky-numbers.html.

The Boy and the Dragon Pearl

Background:

A poor boy and his mother lived near the river, Min, in Sichuan Province. More than one myth is told of dragons and pearls, and they are a significant theme in Chinese dragon art beginning in the Tang dynasty. Some art historians believe that the pearls represented heavenly bodies such as the sun or moon, and that dragons playing with the pearls actually represent dragons attempting to eat the sun or moon, thereby causing an eclipse. (And indeed, some dragons were associated with eclipses. Longwang, for instance, though friendly toward humans, hated the sun so much that he would raise up and try to bite it.) Others say the pearls are a mark of imperial rank, and still others say that they are secretions of the dragons, similar to snake-pearls but far more valuable. In any case, it is clear that they represent items of great value with potentially magical properties.

Another notable aspect of this myth is its emphasis on water, as many of the Chinese stories focus on water or the lack thereof.

Myth:

One myth that is still told is about Xiao Sheng, a boy who loved to sing. He supported his mother and himself by cutting grass and bringing it to the village to sell. It wasn't much, but it was just enough to keep them alive. But through it all, Xiao Sheng would sing, every day.

A horrendous drought came to the area where they lived. There was less and less grass to cut and sell. Still the boy sang.

It happened that this one day, Xiao Sheng came upon a patch of truly beautiful green grass. He quickly cut it all and brought it to the village. He sold it for more money than he could remember. The next day, he returned to that very same patch and, amazingly, the lush green grass had grown back! He thanked whichever gods were bestowing this bounty. Each day, he went back to that patch, and each day, the lush green grass had grown back!

Then, Xiao Sheng had an idea. He decided to dig up some of the grass and plant it near his home. Then he wouldn't have to travel so far each day to cut the grass and bring it to the village for sale. He was very careful about digging up the grass, roots and all. He didn't want to damage any of the plants. Then, when he was almost finished, he noticed that something was underneath the grass. It was a beautiful pearl with a bit of pink coloring. He brought it home to his mother. She wanted to savor the beauty of the pearl for a while, so she put the pearl in their nearly empty rice jar.

When Xiao Sheng went to his patch of grass the next day, he was sad to find that the grass had completely dried up. He ran home full of sadness that he had destroyed the lush green grass.

But then He remembered the pearl and checked the rice jar. He was totally amazed to find that the jar was overflowing with rice! It was then that he and his mother knew they had a magic pearl. So they decided to put it in the box with their very few coins, and soon, the coins were many and they were rich!

They shared their wealth with many, but some people are never satisfied. In particular, two of the men in the village wanted the source of Xiao Sheng's wealth. They went to his hut and demanded the gold and when they saw the pearl, they demanded the pearl. In order to prevent the men from stealing, Xiao Sheng put the pearl in his mouth. The men became violent toward the boy, shaking him and pounding on his back. All of this violence, made the boy swallow the pearl.

Then the boy became extremely thirsty. And he was in pain from this thirst as if his stomach was on fire. His mother gave him drink after drink of water, but it wasn't enough. The boy drank all of the water in the river! Then a very strange and amazing change came over Xiao Sheng. He grew and grew. A thunderstorm erupted and still the boy grew. His eyes grew large. His body grew very large. Horns grew from his head. His hands grew into claws, the claws of a dragon. The storm brought the much needed rain and refilled the river. Xiao Sheng knew he could never change back into a boy. He was a dragon for all time. But he made sure to guard the river and the people.

Illustration:

As with many Chinese dragon myths, this story features a transformation. Here, Xiao Sheng looks over himself after his transformation is complete. The markings on his body are taken from the Asian water monitor lizard, *Varanus salvator*.

SOURCES:

Dekirk, Ash. (2006.) Dragonlore: From the Archives of the Grey School of Wizardry. New Jersey: Career Press, Inc.

Friedman, Amy and Meredith Johnson. (2001, January 28.) The Dragon's Pearl (An Ancient Chinese Legend). UExpress: Tell Me a Story. Retrieved from http://www.uexpress.com/tell-me-a-story/2001/1/28/the-dragons-pearl- an-ancient-chinese.

Ingersoll, Ernest. (1928, 2014.) Dragons and Dragon Lore. New York: Cosimo Classics.

Lawson, Julie. The Dragon's Pearl. Unit 1 No Turning Back. Retrieved from http://ditter91.weebly.com/uploads/5/1/3/7/51371393/the_dragons_pearl.pdf.

McCormick, Kylie. (2012, November 18.) Dragon Pearl. Black Drago. Retrieved from www.blackdrago.com/history/eastpearl.htm.

McLeish, Kenneth. (1996). Myth: Myths and Legends of the World Explored. New York, NY: Faces On File, Inc.

Nickel, Helmut. (1991.) The Dragon and the Pearl. Metropolitan Museum Journal 26, 139-146.

Niles, Doug. (2013.) Dragons: the Myths, Legends, & Lore. Avon, Massachusetts: Adams Media.

Sanders, Tao Tao Liu. (1983.) Dragons, Gods & Spirits from Chinese Mythology. New York: Schocken Books.

The Foolish Dragon

Background:

This myth bears a strong resemblance to one of the stories about Ryujin, the Japanese dragon, and has clear links to Buddhism.

Myth:

A very long time ago, as the myth tells us, there lived a dragon and his wife in the ocean. The dragon noticed that his wife was ailing and became very concerned. He asked his wife if there was anything at all that he could go and get for his wife that would improve her health. She said that she thought eating a monkey's heart would help her health.

So, the dragon went off in search of the heart of a monkey. He went on shore where the tall trees were and saw a monkey. The monkey was eating nuts from the tree. The dragon asked the monkey if he was tired of eating the same old nuts every day. Wouldn't he like a change in his diet? Some variety? From across the sea?

The monkey said, sure, but how would he get there? Thinking that he was being incredible sneaky, the dragon offered to give the monkey a ride on his back. The monkey found himself eating ocean water and became concerned.

"Where are we going?" asked the monkey. The dragon replied honestly that his wife was sick and needed a monkey heart to save herself. The monkey thought quickly and said that that was too bad because he had left his monkey heart back on top of the tree.

The foolish dragon brought the monkey back to his tree. The monkey climbed to the top. He didn't come down.

Some say that Buddha said that at that time, he was the monkey.

Illustration:

The foolish dragon is depicted in this illustration using two important fishes from Chinese rivers: the carp and the catfish. The Mekong giant catfish (*Pangasianodon gigas*) can grow over 8 feet long.

SOURCES:

Latini, David, Ed. (2018.) Epic Tales: Chinese Myths and Tales. London, UK: Flame Tree Publishing.

Sheng, Jim. (2012, June 19.) The Foolish Dragon. Chinese Aesop: Fairy Tales, Folk Tales, Fables, Myths, Legends, and Historical Stories. Retrieved from chineseaesop.blogspot.com/2012/06/foolish-dragon.html.

Wilkinson, Carole. The Foolish Dragon. The Dragon Companion. Retrieved from www.carolewilkinson.com.an/dragoncompanion/thefoolishdragon.php.

Werner, E. T. C. (1922, 1994.) Myths and Legends of China. Dover Publications.

Pearl of Hai Li Bu

Background:

This myth combines several cultural values, including the importance of hunters, loyalty to dragons and the role of dragon pearls.

Myth:

A very, very long time ago, Hai Li Bu was walking along the shore of a stream. Hai Li Bu was known to be a great hunter. As he was walking along the shore of the stream, there was a great commotion in the nearby bushes. It was very loud. When he got close enough, he peered through the bushes and saw a large white goose trying to kill a small snake who was coiled up on a rock. He felt sorry for the little snake. Hai Li Bu was able to make the goose go away and save the life of the snake.

Then an amazing thing happened. The snake grew larger and turned into a woman. The woman was the daughter of the Dragon King. She was very grateful to Hai Li Bu for saving her life. Out of her gratitude, she gave Hai Li Bu a great treasure: a dragon pearl. This pearl was very special. It allowed Hai Li Bu to hear whatever the animals were saying. However, if he revealed what the other animals were saying to humans, he would turn to stone.

Hai Li Bu kept the knowledge to himself for many years. He enjoyed listening to the other animals. However, a day came when Hai Li Bu just could not keep what he heard to himself. The other animals were talking frantically. The birds were taking flight. The animals were saying that there was a great flood coming down the river and everyone who wished to live must run to higher ground as fast as they could. Hai Li Bu ran and told his villagers that they must flee or drown.

Everyone ran to higher ground, including all of the elders and children. Everyone was saved, except Hai Li Bu. He had turned to stone.

Illustration:

The daughter of the Dragon King sits beside Hai Li Bu after he has been turned to stone. Moss grows and birds sit on the shoulders of Hai Li Bu. This dragon is drawn using classical depictions of dragons in chinese art.

SOURCES:

Cherchenuit, Deleios. (2016, April 1). Dragons around the world. TarValon. Retrieved from https://www.tarvalon.net/ content.php?1744-Dragons-around-the-world.

McCormick, Kylie. (2012, November 18.) Dragon History. Dragons of Fame. Retrieved from www.blackdrago.com/history/outline.htm.

Niles, Doug. (2013.) Dragons: the Myths, Legends, & Lore. Avon, Massachusetts: Adams Media.

Candle Dragon

Background:

There are many legends concerned with the Candle Dragon, also known as the Torch dragon in English, or zhu long or zhu yin. Translated, zhu literally means "candle."

The stories tell of a mountain in China, Zhangwei Mountain, north of the Chishui River and far beyond the Northwest Sea. It is a magical mountain.

Myth:

Within Zhangwei Mountain lives the Candle Dragon, a mystical being with a human head and a serpent's body. A huge serpent's body. More than 1000 li (a Chinese mile) long. The body of the serpent which is the body of the dragon is fiery red. His eyes are vertical slits which when closed bring on nighttime. When his eyes are open, it is day time.

This is a very powerful dragon. When he sniffs (think inhale), he brings in summer. When he blows hard, exhaling, it is winter. He never sleeps. In fact, he never eats or drinks. But when he winks, day can turn to night. When he breathes, he can produce torrential rain and gale force winds.

He frequently has a candle in his mouth which can light the way to heaven.

Many say that he represents both dark and light, like ying and yang, which may explain how he still keeps the night and day happening.

Illustration:

The powerful candle dragon has the head of man and the body of a serpent—here, the venomous Russel's viper (*Daboia russelii*) patterning is used. His horns are candles and he also carries one in his mouth.

SOURCES:

Birrell, Anne, tr. (2000.) The Classic of Mountains and Seas. England: Penguin Books.

Chang, Serena. (2015, June 29.) The legend of the Candle Dragon. The World of Chinese. Retrieved from www. theworldofchinese.com/2015/06/the-legend-of-the-candle-dragon/.

Douglas, Lachlan. (2013, October 18.) Chinese Myth of the Candle Dragon. Ancient Chinese Myths and Legends. Retrieved from www.ancientchina7b.weebly.com/1/post/2013/10/chinese-myth-of-the-candle-dragon.html.

The Candle Dragon. (2014.) Cultural China. Retrieved from http://traditions.cultural-china.com/en/13Traditions1298.html.

The Candle Dragon. Myths and Legends. Retrieved from creator.myths.e2bn.org/show/4722.

The Candle Dragon. (2017.) USC Digital Folklore Archives: A database of folklore performances. Retrieved from http:// folklore.usc.edu/?p=36521.

Cultural Questions:

1. Hominins arrive in Asia early in our evolution. Asia is also a source of some of the oldest dragon myths. Compare the myths of China and Southeast Asia. How do they explain power and control of water?
2. Chinese dragons are often benevolent to humans, but sometimes not. Compare the dragons in the Four Rivers of China to Gong Gong.
3. What is the difference between the pearls in Tchang's Pearl and the Boy and the Dragon Pearl?
4. The Candle Dragon appears as one of the most powerful dragons in Chinese culture. He represents both good and evil, light and dark. Discuss these contradictions.
5. In Hai Li Bu, we once more see a pearl, what is the significance in this story of the pearl?
6. How does class status factor into these myths?

CHAPTER 5: JAPAN AND SOUTH KOREA

Background:

Japan and South Korea have many similar dragon legends — some of which are also shared with China — providing an interesting contrast. In Japan, unlike many other Asian cultures, dragons run the gamut from benevolent and self-sacrificing to monstrously evil, with several ambiguous dragons in between. With thousands of yokai (spirits or monsters) in Japanese culture, as well as a history of complex characters in art and literature, the dragons of Japan exhibit a more flexible and nuanced understanding of good and evil inherent to the culture itself.

While human occupation in Japan does not appear as early as in China, archaeology establishes Paleolithic presence more than 35,000 years ago in a warm climate. This was followed by small settled communities near water approximately 10,000 years ago. Water continues to be a prominent factor in the cultural history of the area. One cultural element that is consistent in both Japanese and Chinese history is a patriarchal structure for mainstream society.

SOURCES:

Keally, Charles T. (2009, October 13.) Japanese Paleolithic Period. Retrieved from http://www.t-net.ne.jp/~keally/palaeol. html/

Napier, Susan J. (2005.) Anime from Akira to Howl's Moving Castle: Experiencing Japanese Animation. Palgrave Macmillan.

Ryujin the Sea God

Background:

A Japanese deity for hundreds, perhaps thousands of years, Ryujin has long been held as the ruler of the ocean and many legends have been told about him. The first written account of Ryujin is the Kojiki, from around 712 A.D./C.E., and he is also mentioned in the Nihon Shoki, written around 720 A.D./C.E. One legend held that an emperor of Japan, Empress Jingu, was lent Ryujin's tide-controlling jewels in order to lead an attack on Korea. A woodblock print produced in the 1800s by Utagawa Kuniyoshi depicted yet another myth featuring Ryujin having his tidal gems stolen by a young woman, Tamatora. A form of Shinto known as Ryujin shinko includes the worship of Ryujin and other dragons in association with agricultural rituals. The Watatsumi jinja shrine in Kobe and the Kitakyushu and Daikai jinja shrines in Osaka are all dedicated to Ryujin.

In some tellings of this story it was not Ryujin, but rather his pregnant wife, who was feeling so ill that fateful day. Perhaps this accounts for some of the dragon god's rage. Eventually, Ryujin's wife bore him a daughter, the lovely Otohime, who would become a legend in her own right, as would another of his daughters, Kuniyoshi. He is sometimes said to have transformed into a human to produce Jimmu Tenno, the first Emperor of Japan, though most stories say that Tenno was his grandson or great-grandson through his daughter Otohime.

Ryujin long garnered the fearful respect of sailors, but he was also an ally to Japan, whose rulers he had fathered. During the reign of Empress Jingu, an attack on Korea was undertaken. During the sea battle, Ryujin is said to have turned his tide jewels (or lent them directly to her) so that the water would recede beneath the Korean ships, leaving them stranded. After the Korean soldiers abandoned their ships, Ryujin turned the jewels once more and the water rushed back in, sweeping the soldiers away. Thus Ryujin aided the island nation of Japan in its foreign conquests. Many myths say that he was also instrumental in allying with various political entities in early Japanese history, alternately allying with the Taira samurai class and in particular their leader Taira no Kiyomori, and supporting his rival Go-Shirakawa, the 77th Emperor of Japan.

Myth:

Ryujin, also known as Owatatsumi, was the dragon god of the sea. Ryujin could transform into a human, but spent most of his time living in Ryugu-jo, his castle under the sea made of red and white coral. In this castle, with four great hallways dedicated to each of the four seasons, time passes so slowly that one day in Ryugu-jo is equal to a century on land, though most humans who visited the palace never returned. In this castle he kept his tide jewels, and by turning them he could control all the movements of the sea. Ryujin was a deep blue color, down to his claws, tongue, and horns. His mouth was so wide that whenever he opened it, a whirlpool opened on the surface of the water.

One day long ago, Ryujin was feeling ill and needed to eat the liver of a monkey. He had many servants under the sea - octopi, jellyfish, fish, sea turtles - and so he called for the jellyfish. At this time the jellyfish had bones, and could walk on land. He sent the jellyfish out to collect the monkey liver. And so it was that the jellyfish came upon a monkey in a tree and told him that his liver was needed by Ryujin. The monkey was clever, and knew it could not survive without its liver. He made up a story on the spot, and told the jellyfish that he would be more than happy to give Ryujin his liver, but he would have to go and get it. The monkey said that he knew how valuable his liver was, so he had hidden it in a jar in the forest, but as soon as he found it, he would come to Ryugu-jo himself to deliver it. The jellyfish was quite happy to hear this and waited patiently, until finally he realized that the monkey was not coming back and was forced to return to the sea dragon with the bad news. Ryujin immediately saw how the jellyfish had been duped, and knew that capturing a monkey would be even more difficult now that the jellyfish had let slip why he wanted one. In a fit of rage, Ryujin beat the jellyfish so badly that his bones were shattered, which is why the jellyfish has no bones today.

Illustration:

While few real reptiles are completely blue in color like the Ryujiun of legend, Japan is home to the Japanese five-lined skink (*Plestiodon japonicus*), a species with a bright blue tail meant to distract predators from attacking their head.

SOURCES:

Dalkey, Kara. (2000.) Genpei. New York: Tor Book.

Davis, F. Harland. (1992.) Myths and Legends of Japan. Mineola, N.Y.: Dover Publications.

McLeish, Kenneth. (1996). Myth: Myths and Legends of the World Explored. New York, NY: Faces On File, Inc.

Niles, Doug. (2013.) Dragons: the Myths, Legends, & Lore. Avon, Massachusetts: Adams Media.

Rose, Carol. (2000.) Giants, Monsters, and Dragons: An Encyclopedia of Folklore, Legend, and Myth. New York, USA: W. W. Norton & Company, Inc.

Oto-hime

Background:

Legends about Ryujin's daughter Oto-hime, sometimes known as Toyotama, seem to be as old as those of the dragon king of the sea himself. Oto-hime appears, like her father, in both the Kojiki and the Nihon Shoki. A Japanese scroll dating from the turn of the 17th century depicts another Oto-hime myth, in which fisherman Urashima Taro saved a turtle's life and was taken to Oto-hime's underwater castle. Returning to his village, he finds that others have aged while he has stayed young; seeing that this pains him, Oto-hime gives him a magical box that advances him to his rightful age. Thus, Oto-hime frequently appears as a sympathetic figure in Japanese mythology. The myth of Oto-hime is still celebrated today, and many characters in contemporary anime series are named after her. Scientists also dubbed an extinct crocodile species after her alternative name, calling it Toyotamaphimeia, after this myth.

Myth:

One day Hoori, the handsome son of the sun goddess Amaterasu, went diving into the water to find his brother's favorite fishing hook, which he had lost. While he was searching for the hook, he came upon a beautiful woman, who was actually the daughter of the sea god Ryujin. Her name was Oto-hime, though she is also known by the name Toyotama-hime. Oto-hime asked her father Ryujin if she might help Hoori find the missing fish hook. Ryujin consented, and the two fell in love and were married.

Oto-hime told Hoori about her father, and the two lived in Ryujin's underwater castle for many years. However, Hoori began to grow sad surrounded by water, never feeling the beloved sun on his face. Oto-hime agreed that they should move to land to live for a time.

While the couple was living in a home near the sea, Oto-hime became pregnant. She gave Hoori the order to allow her to give birth alone, and to not set eyes upon any part of this process. Hoori agreed to this, but toward the end of Oto-hime's labor he became curious and peaked through the door. Instead of the beautiful woman he had lived with for so long, he saw a great sea creature writhing in the throes of labor on the floor. With a turn of her body, Oto-hime saw her husband through the door spying on her. She was ashamed of her natural form, but even greater than that, she was heartbroken that her husband had betrayed his word to her. Oto-hime gave birth to a son, Ugaya Fukiaezu no Mikoto, but immediately after she threw herself into the sea and returned back to her father's kingdom, leaving Hoori and their young son behind.

Oto-hime nevertheless could not stand to abandon her son, and so she sent her sister Tamayori to raise him. As the child grew older, some say that he fell in love with his aunt Tamayori and that after they married, they gave birth to Jimmu Tenno, the first Emperor of Japan

Illustration:

In this image the daughter of the great Ryujiun, Oto-hime, is seen giving birth in her true dragon form by her spying husband. Her design is inspired by the red dragonet (*Foetorepus altivelis*), an oceanic species.

SOURCES:

Amika, Toshio. (1993.) The Origins of the Grand Shrine of Ise and the Cult of the Sun Goddess Amaterasu Omikami. Japan Review 4, 141-198.

McLeish, Kenneth. (1996). Myth: Myths and Legends of the World Explored. New York, NY: Faces On File, Inc.

Ō, no Yasumaro. (2014.) The Kojiki : an account of ancient matters. Translated by Gustav Heldt, New York: Columbia University Press.

Benzaiten/Benten

Background:

It is likely that Benzaiten is actually a version of the Hindu goddess Saraswati, who is a goddess of music, arts, wisdom, and learning. She is acknowledged as the goddess of the sea, but moreover, as the goddess of everything that flows, making her the goddess of music, art, literature, and femininity. Saraswati's name means "flowing" and "water." Benzaiten rides atop a great white dragon, and a white serpent — perhaps the very same one — is known to be her messenger. While the white color is unusual for a Japanese dragon, this imagery likely stems from earlier Indian depictions, as Saraswati was depicted dressed in white and sitting atop a white lotus. As stories of the goddess Saraswati came to Japan as early as the 6th century, she came to be known as Benzaiten, another daughter of Ryujin and the deity of love. She often protects humans, especially children. While Saraswati carries a veena of India, Benzaiten carries a biwa of Japan. The legends surrounding Benzaiten may actually predate those of Ryujin, given that Buddhist monks wrote about comets associated with her in 552 and 593 A.D./C.E. and she appears again in the Enoshima Engi, written in 1047 A.D./C.E.

The story of Benzaiten conquering a dragon may have come from a story about Saraswati defeating a three-headed vritra known as Ahi — a great serpent. The island of Enoshima in Sagami Bay built a shrine to Benzaiten which still stands today, one of the three major Benzaiten shrines (along with the Chikubu Island shrine in Lake Biwa and the Itsukushima Island shrine in Seto Inland Sea). There are also Shinto shrines to Benzaiten, where she is more heavily associated with Ryujin.

Myth:

Because she is a daughter of Ryujin, Benzaiten is often said to take the form of a dragon herself, but she is most often depicted riding the white dragon while playing a traditional Japanese instrument, the biwa.

In one story, she appears as a human woman with two sisters beside her, floating above the ocean, to hear the request of a member of the Imperial family, Takakura, for his father Go-Shirakawa's protection and assistance from Ryujin and all his dragons. After Takakura cut open his hand to show his devotion in making such a request, Benzaiten healed his hand with a touch and sent his message to Ryujin, who consented to protect Go-Shirakawa.

However, the most well-known story of Benzaiten shows her role as a benevolent protector. It is said that on the mainland, an evil dragon terrorized the small town of Koshigoe. It liked to feed on the children best. Benzaiten appeared in the clouds on her great white serpent, searching for the dragon. Beneath her, in the water, the island of Enoshima rose up to meet her as she landed. She called the dragon forth, and after speaking with him for a long while, they came to an agreement. Benzaiten would marry the dragon and always be his wife, so long as he reformed his ways and stopped eating humans.

Illustration:

Here Benzaiten rides her great white serpent to protect the town Koshigoe from an evil dragon. Her mount is based on the Japanese eel (*Anguilla japonica*) and the evil dragon is based on japanese depictions of dragons from this period.

SOURCES:

Aldington, Richard and Delano Ames, Transl. (1959.) New Larousse Encyclopedia of Mythology. Hong Kong: Prometheus Press.

Dalkey, Kara. (2000.) Genpei. New York: Tor Book.

Davis, F. Harland. (1992.) Myths and Legends of Japan. Mineola, N.Y.: Dover Publications.

Jobes, Gertrude. (1962.) Dictionary of Mythology Folklore and Symbols. New York, NY: The Scarecrow Press, Inc.

Juhl, R. A., E. G. Mardon, and A. A. Mardon. (2007.) Documentary Evidence for the Apparition of a Comet in Late 593 and Early 594 AD. Lunar and Planetary Science XXXVIII, 1184.

Ludvik, Catherine. (2001.) From Sarasvati to Benzaiten. Ph.D. Thesis, University of Toronto, National Library of Canada.

Williams, George M. (2003). Handbook of Hindu Mythology. Santa Barbara, California: ABC-CLIO, Inc.

Mizuchi

Background:

The word mizuchi refers to a type of aquatic Japanese dragon. The Nihon Shoki text records an entry for the year 379 A.D./C.E., referencing several mizuchi that once lived in Japan including the mizuchi myth written below. A separate mizuchi reference appeared again in the mid to late 700s A.D./C.E. in a collection of Japanese poetry, the Manyoshu. Much later, a 19th century print by Yosai Kikuchi depicts Agatamori slaying a mizuchi with a sword.

Folklorists have debated about the word "mizuchi." "Mizu" means water, and some see the "chi" ending as having an etymology originally meaning "snake," "revenge," or "spirit." Many have argued that mizuchi might be an early form of kappa, water spirits that are generally thought to be harmlessly mischievous but who are sometimes depicted as hungry for human liver. While kappa are usually seen as imps or demons rather than as dragons, the association with mizuchi and with Ryujin's famous propensity for eating primate liver make for interesting historical connections.

Myth:

One mizuchi lived in the Kawashima river, breathing out venom which poisoned many visitors to the river. A man named Agatamori, who would become an ancestor of the Kasa no omi clan, approached the river. He threw three gourds from the calabash tree to the surface. He then taunted the dragon, saying that it could not make the gourds sink and that if he did not, Agatamori would come and kill him for his weakness. The dragon transformed itself into a deer so that he could jump onto the gourds and make them sink, but each time the gourds rose to the surface again. While the mizuchi busied itself with this task, Agatamori slayed it before it could return to its dragon form. However, Agatamori was not done. He knew that the dragon had fellows living in the water below. He spent a long time pursuing each of these dragons until he had found a way to kill each one, and turned the river red with their blood. Afterward, the river Kawashima was known as the pool of Agatamori.

Many years later, in 323 A.D./C.E., it was reported that a dragon like a mizuchi in the Yodo river was causing trouble, breaking through all the dykes set up along the river. The Emperor, based on a dream he had had, declared that two particular men were to be brought and sacrificed for the river dragon. One of the men, perhaps using the story of Agatamori, threw a calabash gourd into the river and asked for it to be sunk if it was the will of divine beings for him to be sacrificed. Although a strong wind came, the gourd did not sink and the man was spared his life.

Illustration:

In the story of Mizuchi, this venomous dragon is perplexed by the buoyant properties of gourds. Here, he is depicted using elements of the japanese pit viper (*Gloydius blomhoffii*), one of the most venomous snakes in Japan.

SOURCES:

Aston, William George. (1905.) Shinto: (the Way of the Gods). London, England: Longmans, Green, and Co.

Bane, Theresa. (2015.) Encyclopedia of Beasts and Monsters in Myth, Legend, and Folklore. Jefferson, North Carolina, USA: McFarland & Company, Inc.

Foster, Michael Dylan. (1998.) The Metamorphosis of the Kappa: Transformation of Folklore to Folklorism in Japan. Asian Folklore Studies 57, 1-24.

JHTI. (2002.) Nihon Shoki," Japanese Historical Text Initiative (JHTI). Berkeley, CA: University of Berkeley, English tr. Aston.

O Goncho

Background:

While the myth of O Goncho was collected and published in 1886, it was likely part of local legend long before then. At that time, locals reported that O Goncho had appeared three times, and since he only appears every 50 years, the myth may have been 150 years old even then.

Despite being at the center of a short myth, O Goncho is nevertheless a famous and feared Japanese dragon. His white color was somewhat unusual for Japanese dragons, but being the traditional color of funerals in Japanese traditional society, may have associated him with the concept of death. Like many dragons, he is linked to a particular location which is still a favorite for Japanese visitors.

Myth:

Near Kyoto, in the Yamashiro province, is a pond so wide and deep that small waves roll across it on windy days. Although the locals swim in the pond, named Ukisima, none swim too far towards its center. There lies the ghostly white dragon O Goncho.

O Goncho sometimes surfaces but is known to quickly swim away once spotted by humans. Though he dislikes being seen, O Goncho will transform every 50 years into a great golden bird and rise above the pond. He will let out a cry like a wolf, and this cry brings with it a terrible famine. Although locals were known to say prayers to keep O Goncho from issuing his cry, they also accorded him great respect, thus putting the signifier "O" in front of his name.

Illustration:

Every 50 years O Goncho will transform from his aquatic state into a great golden bird. This transformation is drawn here using body parts from the carp and crane.

SOURCES:

Bane, Theresa. (2015.) Encyclopedia of Beasts and Monsters in Myth, Legend, and Folklore. Jefferson, North Carolina, USA: McFarland & Company, Inc.

Gould, Charles. (1886.) Mythical Monsters. New York: Cosimo Classics.

Legg, Gerald. (2006.) Dragons. Great Britain: Book House.

Rose, Carol. (2000.) Giants, Monsters, and Dragons: An Encyclopedia of Folklore, Legend, and Myth. New York, USA: W. W. Norton & Company, Inc.

Yamata no Orochi

Background:

The story of Yamata no Orochi comes from two ancient texts, one from Kojiki in 680 A.D./C.E. and the other from the Nihon Shoki in 720 A.D./C.E. Both of these describe our hero Susanoo as the brother of the sun goddess Amaterasu, making him the brother of Oto-hime's husband Hoori. However, other stories place him as one of Ryujin's sons. In both versions, Susanoo is expelled from his home and is roaming the earth when he finds a situation that calls for his supernatural gifts.

While the number 8 figures prominently in this tale, this is not due to any dark associations with the number itself. In fact, the number 8 is considered lucky in both China and Japan, though for different reasons. While one culture may have introduced the idea to the other, in Japan, the kanji for the number 8 consists of two lines splitting away from each other, gradually opening up. This is said to give the appearance of expansion, growth, and prosperity. "Yamata no Orochi" translates into "big snake of 8 branches," and perhaps the association of the number with growth is why the dragon associated with "8" is one of the most massive, expansive Japanese dragons. One might very well make the claim, though, that prosperity without moderation — intake without restraint — was the immense dragon's ultimate downfall.

Myth:

With a name that translates into "big snake of eight branches," Yamata no Orochi was said to have eight heads and eight tails, and was large enough to lie across eight hills and eight valleys. The dragon's sixteen glowing red eyes were constantly looking around so that it could always see what was coming. Its eight tails whipped around producing a sound like whipping wind. For years this dragon lived at the base of Mt. Sentsuzan devouring anything that came near. He loved to eat humans and especially relished the young virgin he demanded once a year from nearby towns. He lay there for so long that his body became covered in moss, and trees began to sprout on his back.

One great family had given seven of their daughters to the dragon as sacrifices, and as the time approached to offer another virgin, their eighth and final daughter Kushinada-hime was chosen. Hearing that she was to be given to the terrible dragon, she ran from her village crying. As it happened, the storm god Susanoo had recently been expelled from Heaven for playing tricks on his sister Amaterasu the sun goddess. Taking the form of man, Susanoo was wandering the world when he heard the cries of a woman. He finally saw the beautiful young lady crying alone, and heard her tragic tale. He could not stand the thought of such a kind, lovely girl being given to a monster. And so, Susanoo escorted the young woman home and asked her grieving parents many questions about the dragon. He then asked for their permission to marry her, if he could slay the dragon and spare their family this great loss. The parents were unsure if he would actually be able to kill the beast, as many had tried before, but the idea gave their hearts hope and they quickly consented.

Suddenly, Susanoo used his magical abilities to transform Kushinada-hime into a comb, which he tucked into his hair to keep her close by. He then asked for eight large vats of the strongest sake that could be found, and for a fence to be built around the family's home. Susanoo laid the vats of sake around the fence and waited for the dragon's arrival.

When the girl was not brought directly to him, Orochi gathered up his massive body and lumbered to the family's home to claim his prize. When he arrived, he instantly smelled the aroma of the sake and each head drank deeply from one of the eight vats, for Orochi was engorged with both food and drink. The wine was so strong that the great dragon collapsed in a stupor. Seeing the dragon passed out, Susanoo approached with a mighty sword and cut off each of the dragon's heads before cutting his body into small pieces. When he began cutting off his tails, he found in the fourth tale an impressive sword, which would be named Kusanagi no Tsurugi and become one of the three Imperial symbols of Japan. Because of Orochi's size, the dismemberment took days and for weeks after the River Hi ran red with the dragon's blood. But the towns around Mt. Sentsuzan all celebrated, because they were finally free of the terrible dragon.

Illustration:

Yamata no Orochi is so large and moves so little that his body becomes covered in moss and trees. This drawing used the Japanese giant salamander (*Andrias japonicus*) as a basis for this polycephalic, many-tailed beast. Real *Andrias* can grow up to 5 feet long and use their many folds of skin to breathe underwater.

SOURCES

Aston, William George, transl. (1972.) Nihongi: Chronicles of Japan from the Earliest Times to A.D. 697. Book I. Clarendon, VT: Tuttle Publishing Co.

Chamberlain, Basil H., transl. (1981.) The Kojiki, Records of Ancient Matters. Clarendon, VT: Tuttle Publishing Co.

McCall, Gerrie. (2007.) Dragons: Fearsome Monsters from Myth and Fiction. New York, NY: Tangerine Press.

Niles, Doug. (2013.) Dragons: the Myths, Legends, & Lore. Avon, Massachusetts: Adams Media.

Ouwehand, Cornelius. (1958-1959.) Some Notes on the God Susa-no. Monumenta Nipponica 14(3⁄4), 384-407.

Kiyo-hime

Background:

The myth of Kiyo-hime is an ancient one from the Wakayama prefecture in Japan. It appears in the ancient text such as the Konjaku Monogatarishu, a collection of tales written in the late Heian period (794 - 1185 CE), as well as in the famous noh play Dōjō-ji and many kabuki plays. Kiyo-hime is also the subject of many Japanese traditional artworks, including an early 19th century netsuke of Kiyo-hime as a snake wrapped around the bell, a woodblock print entitled "Kiyo-hime turning into a serpent" by Yoshitoshi Tsukioka (1890), and a later woodblock by Tadamasu Ueno entitled "Kabuki drama Musume Dojoji - role of the demon of Kiyohime" (1952). The myth mirrors an earlier one, wherein Prince Homuchiwake took a bride for one night but fled when he saw that she transformed into a snake while he slept. Enraged, she chased after him into a lake, much like the spurned lover Kiyo-hime.

While snakes are often abhorred by humans, in Japan they are considered sacred due to their connection with Ryujin. While they are associated with death and thunder, in Shinto temples and households snakes were traditionally not to be harmed, as snakes are seen as a lesser form of dragon.

Transformation stories are also common in Japan, with many powerful deities being able to take animal or human form, and most yokai or spirits are female. The force of Kiyo-hime's sadness and anger are enough to give her the power to transform, but the change is a tragic one that ultimately leads to both her demise and her lover's. While some see in the story a demonization of female desires and emotions, others link her transformation to the Buddhist notion of karma, wherein all emotions and actions result in some eventual transformation of the person.

The story of Kiyo-hime has undergone changes, with some scholars arguing that the Shinto version was designed to emphasize the bad consequences of illicit romances and the danger of seductive women, while the Buddhist version made Kiyo-hime more sympathetic, a widow whose attachment to Anchin stemmed from their love in a previous life. Despite taking place in the year 928 A.D./C.E., the most commonly retold versions of the tale emerge in the fourteenth and fifteenth centuries in Japan. Later Buddhist versions extend the ending, with a priest at Dojoji seeing a vision of a monstrous dragon who told him that he was Anchin himself, married to Kiyo-hime because he had deceived her. Feeling great pity for them, the priest prayed for them and had a copy of the Lotus Sutra created in their honor. Afterward, he received a vision of both of them together, released from their dragon form and in heavenly bliss. This means that even though Anchin was likely taking the Kumano pilgrimage, it was ultimately Buddhism which saved him.

Myth:

Kiyo-hime's father had been blessed with a great house along the Hidaka riverbank. With so many extra rooms, and a great deal of travelers on their way to the shrine at Dojoji, rooms were offered at the great house for a modest price. This is how Kiyo-hime met the young priest Anchin.

Some stories say that Anchin was never interested in Kiyo-hime, or that her father had put the idea of romance between the two of them in her head with a harmless joke, but most agree that Anchin and Kiyo-hime began a passionate affair. Only after securing her heart did Anchin realize that he could never abandon the life of a priest. He set out on his way to the shrine, crossing the river without her while she stood on the shore crying out for him. Finally, she waded into the water after her lover, and through her great despair and anger a great change was wrought in her. Her skin turned to scales and her face became monstrous. Anchin became terribly afraid and ran to the temple, begging the priests there to hide him from her.

When Kiyo-hime arrived, she could not see Anchin. He had been hidden under the temple's large bell. But with her new dragon senses, she could smell him. First she lashed out with her tail, knocking the bell over and over again. When Anchin would not come out to face her, her heartbreak and rage grew again and she unleashed a stream of fire that liquified the bell and burned Anchin inside of it. The twisted piece of metal, holding the remains of the young priest, fell into the Hidaka river never to be seen again.

Illustration:

Kiyo-hime's rage transforms her into a terrible fire-breathing serpent, shown here using elements of the predatory lancetfish (Alepisauridae). The bell hiding her husband is influenced by Japanese design.

SOURCES:

Akinari, Ueda. (2012.) Ugetsu Monogatari or Tales of Moonlight and Rain (Routledge Revivals): A Complete English Version of the Eighteenth-Century Japanese Collection of Tales of the Supernatural. New York: Routledge.

Bathgate, Michael. (2008.) Stranger in the Distance: Pilgrims, Marvels, and the Mapping of the Medieval (Japanese) World. John Hopkins University Press in Medieval Studies. Essays in Medieval Studies 25, 129-144.

Bushell, Raymond. (1977.) Concerning the Walters Collection of Netsuke. The Journal of the Walters Art Gallery 35, 77-85.

Heita's Uwibami

Background:

Like mizuchi, uwibami are a general type of ancient dragon in Japan. While many uwibami are depicted as giant serpents in woodblock prints, their ability to fly and snatch up large animals and humans alike is legendary. The story of a man named Heita's encounter with at least one uwibami were recorded in the 1800s and early 1900s in both texts and woodblock prints.

There are many Japanese woodblock prints of Heita's victory over the uwibami, though the many variations of the story suggest that perhaps Heita killed not just the Settsu uwibami but many, many more in his lifetime. The one by Utagawa Kuniyoshi shows Heita cutting his way out from inside the beast, while Utagawa Kuniteru's woodblock showed yet another version wherein Heita killed the uwibami with a polearm. There are also multiple versions of who Heita actually was - some sources list his name as Yegara-no-Heida Tanenaga, while others list him as Wada Heita Tanegara, and still others list him as Egara no Heita. If he was a historical leader or political figure of some sort, this information has been lost as his warrior status has endured.

Myth:

Many uwibami once plagued Japan. These monsters were giant winged serpents who could swallow a human whole. One uwibami in Settsu province grew so large that it could eat an armored man on horseback and swallow the both of them in one gulp.

Finally, a hunting party was formed to end the beast's reign of terror. One well known warrior, sometimes named Wada Heita Tanenaga but also named as Yegara no Heida, or Egara no Heita Tanenaga of the Wada family, volunteered to join the party. Heita lived during the late 12th and early 13th centuries, and killing the great uwibami is one of many legendary adventures attributed to him.

Most accounts of this hunt describe Heita luring the uwibami into a mountain cave, using the darkness to his advantage to slay him. However, at least one version of the story paints a different picture. The darkness Heita used to his advantage was not the darkness of the cave, but the blackness inside the uwibami itself. Heita allowed himself to be swallowed, then slashed his way through the uwibami from inside of it, until he finally broke free of the great serpent's flesh.

Illustration:

This illustration of Heita's Uwibami is based on the red-crowned crane (*Grus japonensis*) mixed with elements of lizards and deer.

SOURCES:

Dekirk, Ash. (2006.) Dragonlore: From the Archives of the Grey School of Wizardry. New Jersey: Career Press, Inc.

Joly, Henri L. (1908.) Legend in Japanese Art: A Description of Historical Episodes, Legendary Characters, Folk- lore Myths, Religious Symbolism. London: John Lane, The Bodley Head.

Rose, Carol. (2000.) Giants, Monsters, and Dragons: An Encyclopedia of Folklore, Legend, and Myth. New York, USA: W. W. Norton & Company, Inc.

Uwabami. (2019.) Yokai.com. Retrieved from http://yokai.com/uwabami/.

Yofune-nushi

Background:

Although this myth takes place long ago, it was recorded only recently. The first written account was taken in 1918 by Richard Gordon Smith. Due to its early 20th century roots and its author being a visitor to Japan - despite his report that the story came from Japanese locals - opinions vary as to how much the story may have been influenced by Western sources. Still, the tale is compelling and has become a modern favorite despite its less historical roots.

If this story does indeed come from Japanese folktales of the 19th century or earlier, it would have pushed back against several cultural norms. Tokoyo's father being an exiled samurai would be either unlikely or terribly shameful, since samurai who had disgraced their leaders were honor bound to end their lives. Some versions of the story place the samurai in jail on the island town where Tokoyo tracks him, but jail as a punishment for a samurai would be equally unlikely. Tokoyo herself is an unusually active female protagonist, and the pro-feminist message of the story is furthered when she opposes the sacrifice of young girls. However, one can also see the theme of heroes fighting female sacrifice to dragons in the undeniably historical Japanese tale of Yamata no Orochi. While we may never know the extent to which this legend was Japanese in origin or influenced by outside sources, it has been embraced by many Japanese people and dragon myth fans today.

Myth:

Around 1320 A.D./C.E., a samurai named Oribe Shima was banished from his hometown by a local chief. His lovely daughter, Tokoyo, missed her father greatly. The two had been very close, and he had even trained her in many martial art forms. Tokoyo knew that her father had traveled to the Oki Islands, so Tokoyo gathered what money she had and set off after him alone. She finally reached Akasaki, but had so little money that no boatman would take her to the Oki Islands which were in her sight. Tokoyo walked along the beach until she found an abandoned wooden boat, very old and hardly seaworthy. But Tokoyo had grown up diving for pearls and playing in the sea, and she lacked neither skill with a boat nor courage. She piloted her small, wobbly boat to the Oki Islands.

After several days, Tokoyo was still searching for her father when she wandered across a girl dressed all in white, as if for a funeral. The girl was weeping as she looked out over the water. Tokoyo then saw a priest nearby, also dressed in white, and asked what was happening. They explained to her that a terrible dragon named Yofune-nushi controlled the waters off of their small town, which was completely dependent on fishing. If they did not sacrifice a virgin to the dragon every year on June 13th, he would devastate their ships and create hunger and hardship throughout their town. Tokoyo then volunteered to be the sacrifice to Yofune-nushi, shocking both the young girl and the priest. She asked for the girl to hand over her white mourning dress, and sat patiently waiting for the dragon.

Finally she saw a stirring in the water and, placing a dagger between her teeth, dove into the moonlit water. Many silver fish swam around her, but in spite of their frenzied swimming she made out the shape of a huge cave. Swimming to it, she saw that the dragon was not in it but had perhaps gathered things there, for there was a wooden statue she recognized as belonging to the man who had exiled her father. She tucked the status away in her dress, and was contemplating whether she should destroy it or not, when she saw him. Yofune-nushi was 26 feet long, in the shape of a great fish with shimmering translucent scales but with powerful legs. The monster swam toward Tokoyo, and with its great mouth only feet away from her she lunged to the side of it and plunged her dagger into its right eye. The clear water clouded with blood. The beast tried to swim away, but Tokoyo pursued it, slicing its throat. She crawled up to the shore dragging its immense body behind her. The young girl who was meant to be Yofune-nushi's victim ran to the village. The sea monster was then dragged ashore and Tokoyo was celebrated by all.

Tokoyo told and retold her story to priests and villagers for days. And the story spread all the way to the man who had exiled her father and whose wooden statue she had discovered. As it happens, he had been suffering from a mysterious illness which had ended abruptly on the night that the statue had been taken from the dragon's lair. Putting the two events together, he paid his debt to Tokoyo by releasing her father from exile. The two were able to return home together.

Illustration:

Yofune-nushi is an aquatic dragon who does battle with Tokoyo. He is illustrated here as part koi fish, part serpent—his face covered in sensory barbels. He holds a pearl which his source of power, like many other eastern dragons.

SOURCES:

Dekirk, Ash. (2006.) Dragonlore: From the Archives of the Grey School of Wizardry. New Jersey: Career Press, Inc.

Druett, Joan. (2000.) She-Captains: Heroines and Hellions of the Sea. New York: Simon & Schuster.

Smith, Richard Gordon. (1918.) Ancient Tales and Folk-lore of Japan. Montana: Kessinger Publishers.

KOREA

Korea lies between China and Japan. Archaeologically, human habitation seems to be earlier than Japan and later than China. While dragons in China usually have five claws and in Japan have 3 claws, in Korea, dragons usually have four claws. All three areas share the patrilineal, patriarchal dominance in their cultural systems.

SOURCES:

Prehistory and Ancient History. National Museum of Korea. Retrieved from https://www.museum.go.kr/site/eng/showroom/list/760?showroomCode=DM0002.

The Dragon Kings of the Four Seas

Background:

The dragon kings of the four seas were adopted from the Chinese longwang. In Japan, Goko is the dragon of the eastern sea, Gokin is the dragon of the south sea, Gojun is the dragon of the western sea, and Gojun is also the name of the dragon of the northern sea. In Korea, Gwangdeok is the dragon of the east, Gwangli is the dragon of the south, Gwangtaek is the dragon of the west, and Gwangyeon is the dragon of the north. In both cultures, these four dragons were water dragons thought to possess special abilities with the sea much like the four dragons in the Chinese myth about the origins of the four rivers.

Both descend from dragons as well as the political support of dragons have been utilized by Asian nations. In this myth, the four dragon kings side with Korea in a myth that relies on the periodic tsunamis that strike the island nation to "prove" the dragons' support for Korea against Japan. Yet recall that the Japanese Ryujin myths also explicitly state that Ryujin supports Japan's attacks on Korea, even lending his tidal jewels to Empress Jingu for the purpose of defeating the Korean fleet. Each nation would likely vigorously dispute the veracity of the other's claims of draconic assistance. Invoking the support of dragons in political aims acts to provide divine justification for military maneuvers.

Myth:

The legend of the four dragon kings is popular in Korea as well. According to one Korean account, a young man named Samyongtang was sent as an envoy to Japan. Should any trouble befall him, the devoted Buddhist Samyongtang was told that should he face the temple of Hyangsansa to pray, the dragon kings would come to his aid. Furthermore, the dragon king of the western sea had sent a letter urging that he be given aid by his countrymen and that the King of Japan should listen to his arguments against a Japanese invasion.

When Samyongtang arrived in Japan, the King made a marvelous show of providing him with a great and sturdy pavilion, but had plans of setting it on fire with the doors all locked tightly during the night. As the fires began, Samyongtang's friends began collapsing from the smoke and heat. However, Samyongtang remained calm and wrote the symbols for ice to hold in his hands. The fires faltered, and then fizzled out; and slowly icicles began to form on the pavilion. When the king's men went to confirm Samyongtang's death, he calmly inquired about finding less frigid lodgings. When the king tried to burn him again, presenting him with a horse that was actually smouldering iron, Samyongtang turned toward the temple of Hyangsansa and prayed.

Back in Korea, his master heard his cries and dipped his fingers in water. He flicked the water three times, and three colorful clouds emerged beside him. These clouds then began to move swiftly, followed by all four dragon kings. As they approached Japan, the skies overhead darkened, and the ocean swelled. Soon all of the land in Japan was blanketed in water. The king sent a letter of peace, but Samyongtang saw that it still lacked respect. The two spoke at length, the Japanese countryside falling more underwater by the hour, before Japan finally declared brotherhood with Korea - even naming Korea as the older brother - and peace was achieved between the two nations.

Illustration:

Here the dragon kings of the four seas are shown following Samyongtang on his way to Japan. Their patterns and features are serpentine, with some elements from common fishes from the Sea of Japan and Korean river drainages (Korean mackerel, Pacific herring, northern snakehead, Alaskan pollock).

SOURCES:

Transactions of the Asiatic Society of Japan, Volumes 17-18. The Gospel in All Lands.

Gould, Charles. (1886.) Mythical Monsters. New York: Cosimo Classics.

Yongwang

Background:

Yongwang bears a striking resemblance to Ryujin, the sea god of Japan, and it is likely that one or both influenced the other. Yongwang also has a kingdom under the sea and is especially sacred to fisherman and farmers. "Yong" is the Korean word for dragon, while "rong" is the Vietnamese word and "long" is the Chinese word. Like the Korean Yongwang, the Chinese Longwang was associated with water and could adapt to small bodies such as wells.

There are many rituals dedicated to Yongwang, including pungeoje (a ritual to bring a big catch to village fishermen) and yongwangmeogigi (a household ritual for peace and prosperity). Yongwang may also refer to all four of the dragon kings at once - Gwangdeok of the east, Gwangli of the south, Gwangtaek of the west, and Gwangyeon of the north.

Myth:

Yongwang is the ruler of the seas, a water dragon who can control ocean currents, the wind, the waves, the tides, and even rainfall. He is not only important to fishermen but also to farmers because of his ability to make infertile land fruitful again. Across Korea, there are many wells and springs that are thought to be visited by Yongwang when he is not residing in his jewel-filled palace under the sea, Yonggung. Fishing villages often have shrines to Yongwang, where offerings are made to keep sailors safe and bring rain in times of drought. He is revered by both Buddhists and those who adhere to traditional folk beliefs.

Illustration:

Yongwang is said to visit the springs and streams of Korea, the same places that the Chinese soft-shelled turtle, (*Pelodiscus sinensis*) can be found. The shell of the turtle is elongated here into sections reminiscent of another oceanic creature, the chiton (Polyplacophora).

SOURCES:

Bane, Theresa. (2015.) Encyclopedia of Beasts and Monsters in Myth, Legend, and Folklore. Jefferson, North Carolina, USA: McFarland & Company, Inc.

Grayson, James H. (2013.) Korea - A Religious History. New York: Routledge.

McLeish, Kenneth. (1996). Myth: Myths and Legends of the World Explored. New York, NY: Faces On File, Inc.

Cultural Questions:

1. Why is the sea-god Ryujin so heavily associated with Japanese royalty?
2. Compare this story to the Chinese myth "The Foolish Dragon". What features do they share? How do they differ? What contact theories explain the similarity of these myths? Why was the foolish behavior shifted to the jellyfish in the Japanese version?
3. What role does the monkey play in Japanese mythology?
4. How do natural elements feature into these myths? How is nature viewed as sets of distinct kingdoms (such as the sun kingdom and the sea kingdom)?
5. What role does trust play in these tales?
6. How are women viewed in these stories? Do their roles appear to change in different time periods?
7. How are these legends connected to other legendary Japanese spirits and creatures?
8. Japanese society has often emphasized community service over individual desires. Do these stories warn against selfish behavior?
9. What role does transformation play in these stories?
10. Why do the Korean accounts of the dragons of the four seas stand in direct political opposition to Japan and the Japanese stories about these dragons?

CHAPTER 6: OCEANIA AND AUSTRALIA

Background:

Australia is a separate continent which became inhabited by human beings somewhere between 65,000 and 40,000 years ago. The geology and the cultural systems are dynamic.At times, New Guinea and Tasmania were connected to Australia, but not to the mainland of Asia.

In order to understand the aboriginal mythologies of Australia, we need to know something about the various Indigenous cultures that evolved there. Aboriginal cultures focused on the land, and caring for it sustainably. It is the center of their belief system. This means that Australian Indigenous people knew every part of their land, the resources that it could provide, when and where they were available and the appropriate technology to acquire those resources.

The Dreaming, or the Dreamtime, combines spirituality and the sacredness of the land. The Dreaming is a timeless cyclical interconnected system of beliefs, stories, and practices which tie together the spiritual and physical worlds including the origins of life. The Dreaming can hold different meanings for different Aboriginal peoples - and there were possibly 600 communities at the time of European contact - but is generally seen as a living force connecting the land to the people, all of their ancestors and their creation.

SOURCES:

Kamminga, Johan. (1999.) Prehistory of Australia. Australia: Allen and Unwin Pub.

Sanday, Peggy Reeves. (2007.) Aboriginal Paintings of the Wolfe Creek Crater: Track of the Rainbow Serpent. Philadelphia, Pennsylvania: University of Pennsylvania Museum of Archaeology and Anthropology.

Great Rainbow Serpents of Australia

Background:

There are many versions of the Rainbow Serpent story, but all are told through the understanding of the Dreaming. There are a multitude of Indigenous tribal communities, and each have their own stories and names for the Rainbow Serpents, with some stories being at least 4,000 to 6,000 years old. In some stories the Rainbow Serpent is female, others male, and yet others a being with no gender or androgynous. Sometimes the serpent is benevolent, sometimes cruel. Rainbow Serpents are said to live in watering holes and are definitively associated with water. In many groups, they were associated with quartz crystals or mother of pearl and conferred with medicine men, who were thought by many to be the only ones who could safely approach a Rainbow Serpent's watering hole home.

A study of Rainbow Serpent rock paintings from Arnhem Land, located in Australia's Northern Territory, showed that early paintings matched most closely with the Ribboned pipefish, *Haliichthys taeniophorus*, a type of seahorse. Given the ability of male pipefish to carry eggs, this could have contributed to the androgynous nature of the Rainbow Serpent; in the Kimberley people's stories, for instance, the first Ungud is gendered as masculine but lays eggs. During the Pleistocene-Holocene transition, Arnhem Land Indigenous peoples would have seen washed up seahorses, an increase of snakes in their lands, and more rainbows. Over time, it seems that drawings of Rainbow Serpents incorporated more aspects of other animals like crocodiles and macropods, and the contemporary versions have the most varied characteristics borrowed from other creatures.

In modern times, the Rainbow Serpent may be invoked as an anti-colonial symbol. For instance, in 2006, when an international zinc mining operation threatened to disrupt local animal habitats and Indigenous fishing grounds in the Northern Territory, the Gudanji protested under the belief that the local McArthur river is home to a Rainbow Serpent who would be angered and retaliate with all manner of storms. Also in the Northern Territory, in Arnhem Land, the Jawoyn Association use the Rainbow Serpent who has been said to live in their Nitmiluk territory on all their official letterhead and reports, asserting their cultural identity in concert with the landscape.

Because of the ubiquity of the Rainbow Serpent across Australian Indigenous peoples, it can act as a unifying shared cultural symbol for many diverse Indigenous communities. The following myths are a few examples of the many Rainbow Serpent stories in Australia.

Myth:

The Creation Serpent: In the Dreaming, the earth was dark and flat, and all its creatures slept beneath the ground. There, the great Rainbow Serpent Bolong (or Bolung) also slept, but in time she awoke and crawled along the land, her movements carved valleys and built mountains. In some stories, the land and the animals sleep inside the Rainbow serpent's belly, and she spewed them out to create the world. She created the sun, and the colors, and told all of the other animals to wake up from their sleep. The Jawoyn people tell of how the Rainbow Serpent watched the frogs plod along, their bellies heavy with water. She tickled their bellies, and the waters of the world flooded out of their mouths, filling the gorges the Rainbow Serpent had made. Soon after, the plants of the world began to grow.

The Rainbow Serpent told the inhabitants of the world that there were laws they should all follow. Those that followed the laws would be granted human form, those that disobeyed the laws would become stone. This came to pass, and the tribes who were granted humanity were each given a different totem animal by the Rainbow Serpent, which they were forbidden from slaying. This ensured that there was always enough food to be had by all. The Rainbow Serpent became known as the protector of the land and the giver of life. She would punish those who disobeyed the laws. In some legends, the Rainbow Serpent brings the rains, and sometimes causes floods to punish evildoers. It is said that the Rainbow Serpent likes to inhabit deep pools of water, and when she moves between them, a rainbow is seen in the sky. This marks the beginning of the rainy season.

Wollunqua: The Warramunga tribe of northern Central Australia has a strong relationship with the Rainbow Serpent Wollunqua. There is a huge waterhole located in the Murchison Ranges that purportedly never dries up even during the most terrible droughts. This is because the mighty Wollunqua calls that watering hole his home, and his tail is affixed to it. Wollunqua is said to be the sire of all snakes, and is never mentioned by name since doing so gives him even more power. He is at least 150 miles long, and it is said that should he ever stand up, his head would reach the heavens. Wollunqua's job is to protect the watering hole and punish those who drink from it unlawfully. The Warramunga tribe has many ceremonies made to appease the serpent, to prevent it from leaving the watering hole or eating people. They still perform sacred ceremonies to Wollunqua.

Ungud: The Kimberley peoples tell not of a single Rainbow Serpent, but many. For some, Ungud is the name of a singular Rainbow Serpent, who can grant favors such as rainfall, herd increases, and releasing spirits of babies to enable their birth, if certain paintings of him are retouched. In some legends, he rose from the salt water, created land, and laid eggs across the land, which hatched and were called new Ungud(s) or Wandjinas. This may explain why in many stories, the Ungud were a vast number of mythic snake beings who wandered the early earth and shaped the lands. They taught the first people how to make tools, weapons, and laws. After their job was done, some crawled back into the earth, and in these places limitless waterholes can be found, often in the shape of a snake. Others ascended to the heavens, and produce the fertilizing rains upon which all life depends. Every year, the aboriginal people restore the ancient paintings of the Ungud at the beginning of the rainy season, to appease the serpents to give plentiful rain that year. The Ungud are also associated with the fertility of human men.

Illustration:

This image of the Great Rainbow Serpent is based on the largest Australian snake, the scrub python (Morelia amethistina). Growing up to 19 feet long, this snake has iridescent scales which shine different colors in the sunlight. Scrub pythons live in humid forest environments.

SOURCES:

Buchler, Ira R., et al. (2011.) The Rainbow Serpent: A Chromatic Piece. Berlin, Germany: DeGruyter Pub.

Capell, A. (1960.) Language and World View in the Northern Kimberley, Western Australia. Southwestern Journal of Anthropology 16(1), 1-14.

Dreamtime Stories - the Rainbow Serpent. (2001, September 26.) Australia Lesson Activities. Retrieved from http://www.expedition360.com/australia_lessons_literacy/2001/09/dreamtime_stories_the_rainbow.html.

Elkin, Z. P. (1930.) The Rainbow-Serpent Myth in North-West Australia. Oceania 1(3), 349-352.

Gibson, Chris. (1997.) 'Nitmiluk': Song-sites and Strategies for Aboriginal Empowerment. Land and Identity: Proceedings of the Nineteenth Annual Conference; Journal of the Association for the Study of Australian Literature, University of New England, Armidale, September 27–30, Sydney: University of New England Press: 161-167.

Kaberry, Phyllis M. (1936.) Spirit-Children and Spirit-Centres of the North Kimberley Division, West Australia. Oceania 6(4), 392-400.

Morris, Desmond and Ramona. (1965.) Men and Snakes. New York: McGraw Hill Book Co.

Tacon, Paul, Meredith Wilson and Christopher Chippindale. (1996) Birth of the Rainbow Serpent in Arnhem land rock art and oral history. Archaeology in Oceania 31(3), 103-124.

Radcliffe-Brown, A. R. (1926.) The Rainbow-Serpent Myth of Australia. The Journal of the Royal Anthropological Institute of Great Britain and Ireland 56, 19-25.

Radcliffe-Brown, A. B. (1930.) The Rainbow-Serpent Myth in South-East Australia. Oceania 1(3), 342-347.

Stop digging Down Under? A zinc mine in Australia meets resistance among Aborigines concerned about the environment and a 'rainbow serpent.'. (2006). Christian Science Monitor 16, 4.

The Dreaming. (2015, March 31.) Australian Government. Retrieved from http://www.australia.gov.au/about-australia/ australian-story/dreaming.

The Jawoyn People. Nitmiluk Tours: Jawoyn 'Sharing Our Country.' Retrieved from https://www.nitmiluktours.com.au/ about-us/jawoyn-people.

Worms, E. A. (1955.) Contemporary and Prehistoric Rock Paintings in Central and Northern North Kimberley. Anthropos 50(4/6), 546-566.

Hawaiian Mo'o

Background:

124-1025 A.D./C.E. is the estimated date for the first arrival of Polynesians from the South Pacific, to the Hawaiian Islands. They probably built and used double-hulled voyaging canoes. They were experts at navigating the Pacific Ocean, using the stars, ocean currents, and the sun to guide their long voyages. Evidence of sweet potatoes suggest a South American connection. A stratified class system developed based on kinship relations.

Around 1500 years ago, Polynesian explorers sailed across the Pacific ocean and made landing on the Hawaiian Islands. The tales of the Mo'o are ancient, and it is thought that the stories came with the first people of Hawai'i on their boats. Historically, Hawai'i had no native terrestrial reptile or amphibian species, and no fossil reptiles or amphibians have been found there. However, Polynesia is home to a great diversity of herpetofauna, including monitors, geckos, skinks, and agamids. Since the 1700s, several lizard species have been introduced and have had a negative impact on native Hawaiian organisms.

The Mo'o are described in Hawaiian mythology as supernatural, lizard-like beings who often act as guardian spirits. In native Polynesian and Hawaiian languages, the word Mo'o literally means "lizard" or "gecko," and small lizards can be seen as one form of the larger mo'o. The Mo'o of legend are often female, and have the power to shapeshift - often from human to lizard, and from small lizards to massive ones. They can be good or evil, depending on the mo'o. Mo'o are often found in pools of water or rivers, which they protect. In spite of being related to some mo'o, Pele, the goddess of volcanic fire, seems to have a natural dislike for them possibly because of their strong association with water.

According to 19th century historian Samuel Kamakau, the Mo'o would come to the surface when altar fires were lit. The beasts were supposedly 12 to 30 feet long and black in color. His account described their affinity for the intoxicating drink 'awa, and their rocking behavior after consuming it.

Mo'o tales still exist and some are seen as protectors. For instance, a guardian mo'o on the east end of Molokai was said to have been outraged at a developer's attempts to interfere with his sacred pond. In response, he emerged in our dimension and crushed their bulldozers, and caused great distress to one construction worker who saw him.

Here are a few of the many Mo'o tales told on the Hawaiian Islands.

Kalamainu'u

Since Mo'o love the water, there are a great deal of stories where a mo'o helps fishermen and sailors. For example, there is the tale of the Mo'o Kalamainu'u. She would take the form of a beautiful maiden and surf along the beach. Eventually, she fell in love with a young chief of Kauai and they were happily married. The chief longed to return to the surf. Kalamainu'u gave him her surfboard so he could go. Unfortunately, the chief was intercepted by Kalamainu'u's two sisters who were jealous of their sister's happiness. They tried to scare the chief off by telling him of his wife's Mo'o nature, but he didn't believe them. He returned in secret to his house and saw his wife in her true Mo'o form. She spotted him and became angry with him. She attacked.

Because the chief showed no fear at her terrible form, Kalamainu'u forgave him for spying. Then they chased after the two conniving sisters who turned into fish. They attempted to escape, but Kalamainu'u set a basket trap for them. The sisters were caught as fish. According to the story, this is how the trapping technique for hinalea (wrasse fish) was invented. Even today, fishermen ask the spirit of Kalamainu'u for help when catching wrasse.

In some versions of this story, Kalamainu'u allowed her husband, the chief Makea or Puna'aikoa'e, to go to the shore for only a day. There he met her brother Hinale. It was Hinale who told the chief of Kalamainu's true form, and instructed him to spy on her. Hinale also gave the chief an idea for how to keep her away so that he could escape for good - to ask her to bring him chilled water from a far-away mountain. As she tried to fill a gourd her husband had punctured, he escaped and was defended from Kalamainu'u's wrath by Pele. Hinale, attempting to escape his angry sister, jumped into the water and became a wrasse fish. Kalamainu'u only caught her brother with the help of 'Ounauna, the hermit crab, who taught her how to make a trap.

Illustration:

Hawaii has no native terrestrial lizard or snake species, so Kalamainu'u is based on the tokay gecko (Gekko gekko) from Asia and the Pacific islands the Hawaiian people most likely came from. One could imagine how useful the adhesive toepads of geckos would be for surfing!

The Dragon Slayer Hi'iaka:

Not every Mo'o is benevolent, however. In the legend of the two sisters, Pele and Hi'iaka, the Mo'o are seen as dragons to be slain by the hero. In the story, Hi'iaka is the younger sister of the volcano goddess Pele. In a dream, Pele fell in love with the handsome prince of Kaua'i, Lohiau. She awoke, and demanded the prince be brought back to her. Hi'iaka, though afraid of the dangerous journey, accepted, in order to help her sister. On her journey, she conquered many trials. During her passage through a dangerous forest, she was besieged by the shape-shifting Mo'o Pana'ewa, who took the forms of benign objects such as moss and rain so that she might kill Hi'iaka. With the help of the gods, Hi'iaka used her magical pā'ū, a Hawaiian feather skirt, to kill the shapeshifting Mo'o.

As Hi'iaka grew stronger on her travels, she slew dozens more of the tricky Mo'o. The greatest of these Mo'o was Mokoli'i, who appeared before her as a giant sea dragon, blocking the watery passage to Kua-loa. They did battle, Mokoli'i spraying water and vaulting, but she was no match for the heroic Hi'iaka. Hi'iaka erected her severed fluke upward in the ocean, and this became known as the modern island of Mokoli'i. Following this, Hi'iaka discovered that the prince had died in the time it took for her to reach him. This was no problem for Hi'iaka, who simply grabbed his soul from the sky and put it back into his body through his eye socket. When she returned the prince to Pele, Pele rejected the offer - some say she believed that they were having an affair and spewed lava at them - and Hi'iaka took the prince for herself.

Illustration:

The dragon from the Hi'iaka mythos is influenced by the black tipped reef shark (Carcharhinus melanopterus) native to the Hawaiian islands, and the New Caledonian giant gecko (*Rhacodactylus leachianus*).

SOURCES:

Beckwick, Martha. (1970.) Hawaiian Mythology. Honolulu, Hawaii: University of Hawaii Press.

Emerson, Nathaniel B. Pele and Hiiaka. Retrieved from https://archive.org/stream/pelehiiakamythfr00emeriala/ pelehiiakamythfr00emeriala_djvu.txt.

Hawai'i's Mo'o: Tiny Gecko, Seductive Woman, or Water Dragon? CryptoVille. Retrieved from http://visitcryptoville.com/2015/09/30/hawaiis-moo-tiny-gecko-seductive-woman-or-water-dragon/.

Kaleoikapolialoha. (1998.) Tales from the Enchanted Isles Hawaii. Honolulu, Hawaii: Ka 'imi Pono Press.

Manu, Moke. (2006.) Hawaiian Fishing Traditions. Honolulu, Hawaii: Kalamaku Press.

McLeish, Kenneth. (1996). Myth: Myths and Legends of the World Explored. New York, NY: Faces On File, Inc.

Mo'o: Dragons of Hawai'i. (2013, January 10.) Uncanny Hawaii: The Unconventional Guide to Hawaii. Retrieved from http:// uncannyhawaii.com/mo-o-dragons-of-hawaii-vampire-drake/.

Wianecki, Shannon. The Sacred Spine. Maui Magazine. Retrieved from http://mauimagazine.net/the-sacred-spine/.

Melanesia and the Solomon Islands

Walutahanga

Background:

Melanesia and the Solomon Islands are part of a cultural system that is thought to date back to when humans first entered this area of the Pacific Ocean. After contact, Europeans recognized that the peoples in this area have a distinct cultural system, often reflecting the dynamic geographical space that they live in which is a double chain of volcanic islands. The story of Walutahanga has several versions, as well as multiple similarities to some Agunua stories.

Myth:

The story of Walutahanga hails from the Solomon Islands. The story begins with an ordinary woman who soon expected a baby. Unfortunately, she did not give birth to a human baby, but to a serpent named Walutahanga. The mother, fearing harm would come to her monstrous child, hid it in the forest. Despite her efforts, her husband discovered the truth and cut the snake-child into 8 pieces. Following this, there were 8 days of torrential rain. The rain helped the eight pieces of Walutahanga reform into a larger, more dangerous sea serpent.

The serpent sought revenge on the island, and attacked its people in anger. After several days, the people of the island caught and slew Walutahanga. They cooked her into a stew and ate her body. They cast the leftover bones into the ocean. Only a single mother and her daughter did not eat of the stew. Again, it rained for 8 consecutive days and the bones of Walutahanga reformed at the bottom of the ocean into an even more deadly monster. On the eighth day of rain, Walutahanga sent out 8 tidal waves to destroy the island. The only people who were spared were the mother and daughter who had not eaten the stew.

Walutahanga then became the guardian of the island and its people. She gave the survivors the gifts of clean water and coconut trees and she is still worshipped as a protector spirit. The word Walutahanga means ‘eight fathoms.’

In some stories, the mother hid Walutahanga in their very home, where the snake helps care for a second daughter. The father was startled by her one day and cut her into 8 pieces, and she reconstituted herself with 8 days of rain. She wandered for a long time before finding a new home, but became a man-eater living in a cave. She was captured with the help of two man-eating dogs, cut again into 8 pieces, and all in the village cooked and ate her except for one woman and one child. Seeing that they pitied her, she asked that they not eat her head. The villagers then threw the bones into the sea, where they reformed on the bottom. Again, it took 8 days for Walutahanga to reform, and when she did, the sea cracked with a noise like great thunder. This was followed by 8 tidal waves that capsized boats and destroyed the village. The one woman and child were saved, and Walutahanga gifted them with clear streams, pigs, coconuts, yam, and taro. Leaving the villagers behind her, Walutahanga encountered on the water a kindly fisherman, who allowed her to rest her head on his boat and helped her find a new home. She assisted him from that day forward, and was celebrated by many.

Illustration:

The ghostly Walutahanga is drawn here using the skeleton of the Solomon Island skink (*Corucia zebrata*), the largest skink in the world. An arboreal species, they live in family groups and provide parental care for their young which are born through ovovipipary (pseudo live birth).

SOURCES:

Bane, Theresa. (2015.) Encyclopedia of Beasts and Monsters in Myth, Legend, and Folklore. Jefferson, North Carolina, USA: McFarland & Company, Inc.

Fox, C.E. and Drew, F. H. (1915.) Beliefs and Tales of San Cristoval (Solomon Islands). The Journal of the Royal Anthropological Institute of Great Britain and Ireland 45, 131-185.

McCormick, Kylie. (2013, September 9.) Walutahanga. Dragons of Fame. Retrieved from www.blackdrago.com/fame/ walutahanga.htm.

Poignant, Roslyn. (1967.) Oceanic Mythology: The Myths of Polynesia, Micronesia, Melanesia, Australia. London: Paul Hamlyn.

Melanesia

Agunua

Background:

Melanesia has been affected by multiple waves of Oceanic migration, leading to a great diversity across the island's many communities. One common belief is in figonas, mythical creator beings, with slight variations between communities. All figona appear to be serpentine in nature, and were positive figures - little evidence exists for stories of evil figona. Figona could be male or female, or have androgynous properties. Hatuibwari, for instance, was a male figona who had a serpent's body, human head, four eyes, and four breasts, and suckled the creatures he created.

Different Melanesian communities had different versions of Agunua, but she could also be said to fully encompass all of them. Some stories differentiate the local versions of Agunua from the main Agunua, and gender the local versions female and the all-encompassing version male.

With the introduction of Christianity, the cults of Agunua were greatly decreased. Associating the serpent-formed figona with the Christian Devil, despite their overwhelmingly positive nature in the Melanesian myths, caused a fervor against the traditional stories and made some Indigenous converts to Christianity feel shame over their ancestral belief system.

Myth:

In Melanesia, especially the Cristobal district, stories are told of a creator being, the figona. Agunua is known as the greatest figona. Agunua is credited with creating the sea and land, weather systems, animals, humans, and all plant life. The stories say that she created fruits and vegetables to feed the people, but that she had a brother who burned some of the plants and those plants were forever inedible. She created a male child, but it wasn't able to take care of itself so she created a woman "to make fire, cook and weed the garden" (high status activities). The first fruits of a harvest were offered in many villages to Agunua, as is a libation from the first drinking coconut off of a coconut tree. Prayers also call on Agunua to take away sickness and bring health and fortune, especially in relation to gardening, cooking, and building. After death, it was thought that all souls would join with Agunua.

One story about a local form of Agunua, called Kagauraha, is nearly identical to the story of Walutahanga. In it, Kagauraha was babysitting her daughter's son, who began to cry. The father entered and was alarmed, thinking the giant snake was killing his son. He chopped her into pieces, but those pieces began to reunite. Kagauraha, after putting herself back together, left and warned that the crops would fail without her, which they did. On the move, she met two boys who were kind to her and helped her build a new home. In the Banks Island version, the husband does not chop her up but waits until she is alone and sets the home on fire. Dying, she asked her daughter to bury her head, and the first coconut tree sprang from it. This explains the sacredness of the coconut milk, and also seems connected to the Philippine stories of Ulilang Kaluluwa and Galang Kaluluwa.

Illustration:

Since Agunua is often associated with coconut offerings, she is shown here examining some not-yet-ripe coconuts on a palm tree overhanging the ocean. Her design is based off a genus of boas found on the Solomon Islands, *Candoia*.

SOURCES:

Bane, Theresa. (2015.) Encyclopedia of Beasts and Monsters in Myth, Legend, and Folklore. Jefferson, North Carolina, USA: McFarland & Company, Inc.

Bellwood, Peter (1987). The Polynesians – Prehistory of an Island People. Thames and Hudson.

Fox, C.E. and Drew, F. H. (1915.) Beliefs and Tales of San Cristoval (Solomon Islands). The Journal of the Royal Anthropological Institute of Great Britain and Ireland 45, 131-185.

Ivens, Walter. (1934.) The Diversity of Culture in Melanesia. The Journal of the Royal Anthropological Institute of Great Britain and Ireland 64, 45-56.

McCormick, Kylie. (2017, October 9.) Aganua. Dragons of Fame. Retrieved from http://www.blackdrago.com/fame/agunua.htm.

Poignant, Roslyn. (1967.) Oceanic Mythology: The Myths of Polynesia, Micronesia, Melanesia, Australia. London: Paul Hamlyn.

Resture, Jane. (2009, October 7.) Melanesian Mythology: Solomon Islands. Jane's Oceania Home Page. Retrieved from http://www.janeresture.com/melanesia_myths/solomons.htm.

Rose, Carol. (2000.) Giants, Monsters, and Dragons: An Encyclopedia of Folklore, Legend, and Myth. New York, USA: W. W. Norton & Company, Inc.

Mythic Hawaii. Ancient Hawaiian History Origins of the Hawaiian Islands, Culture and Natives. Retrieved from http:// mythichawaii.com/hawaiian-history-culture.htm.

Maori, New Zealand

Taniwha

Background:

Maori traditions tell stories of fantastical sea dragons, the Taniwha, who can be helpful or harmful to humans. The mo'o traditions in Hawaii are similar to stories told in New Zealand. In fact, another word for both taniwha and lizards is "moko." The first people to settle on both island systems were Polynesians which may account for the similarity in the myths. There is no evidence of settlement on New Zealand before 1000 A.D./C.E. Maori oral history tells the story of the first human groups to arrive in what we call New Zealand, but Maori call the land Aotearoa and themselves the tangata whenua.

The taniwha, like the Mo'o of Hawaii, are an entire category of supernatural creatures, and cannot be encompassed by a single story. The taniwha myths come from the Maori peoples of New Zealand, where they are described as magical, reptilian beasts associated with the water. Sometimes they are described more like sharks and whales, other times like winged lizards and tuatara - sometimes they can shapeshift. Taniwha like to inhabit deep pools, mysterious caves, and areas of turbulence in the ocean. They are usually guardian spirits, protecting either a body of water or a group of people. But they can also be malevolent—causing natural disasters and kidnapping people who get in their way. Consider a few examples of their physical descriptions and dispositions: Tutangatakino, a flying enormous lizard who guided the Aotea canoe from Hawaiki and was said to be ridden by Tuhaepo; Haumapuhia, a woman who transformed into a taniwha after being drowned by her father and subsequently changed the riverscape; Tutaeporoporo, a pet shark who gradually turned into a flying taniwha and only eats the Whanganui Maori as revenge for killing his master; and the Lyell Stream taniwha, man-eater yet pet to a chief, who had a scaly body, long neck, ejected poison that knocked victims unconscious, and upon his death was discovered to have treasures in his stomach.

Recently, taniwha have posed as a warning against environmentally harmful developments. For instance, multiple mishaps and deaths associated with the establishment of the Hydro works at Lake Takapo were said to be caused by a taniwha who had been sighted there regularly even into the 1800s, by Maori and non-Maori alike. However, some argue that this creature is a good guardian spirit, a tipua, and that taniwha are generally bad.

Here are a few taniwha myths that exemplify their diversity.

Arai-te-uru

Myth:

According to legend, the ancestors of the Maori peoples travelled to New Zealand from the land of Hawaiki, guided by several taniwha. The most powerful of these old taniwha was named Arai-te-uru. There was little trouble on the journey thanks to the taniwha's protection, though the people had to help them when they became caught in nets along the way. When they came close to the land, Arai-te-uru made sure to smooth out the path so that the ships could land safely. Some say a fellow taniwha, Rua-mano, accompanied her as they oversaw the Takitimu canos.

The ships made landing in what is known today as Hokianga Harbour, located at the northern tip of the country. Very shortly after landing, Arai-te-uru gave birth to eleven taniwha sons. The sons loved to dig, and the oldest of the sons, Waihou, claimed he was the best at burrowing. Arai-te-uru heard this and told her sons to dig as far as they could and come back to tell her what they had found in this new land. The displacement of land made from the taniwha burrows can still be seen in the geography of Hokianga Harbour. Arai-te-uru and another taniwha Niua (possibly one of her sons) still live in the caves of Hokianga Harbour, where they act as guardians for the people there. The area is protected as the Araiteuru Recreation Reserve. "Arai-te-uru" is also the name of a canoe that voyaged to New Zealand.

A similar story tells of the crew of the Aotea canoe, and how they were escorted from Hawaiki to Aotearoa (New Zealand) by a flying taniwha named Tu-tangata-kino. She took up residence along the Whanganui River, accepting offerings from those who canoed by her territory. A powerful local leader, Tuhaepo, was known to occasionally ride this taniwha.

Illustration:

Here, the mother Taniwha lays with her clutch of 11 eggs. She is inspired by the tuatara (*Sphenodon punctatus*), which is not a lizard but the only surviving member of an ancient order of reptiles called Rhynchocephalians. They are only found in New Zealand. Her wings are based on those of long tailed bat (*Chalinolobus tuberculatus*).

Horomatangi

Myth:

It is said that Horomatangi was brought from Hawaiki by Ngatoro-i-rangi, a great explorer. Horomatangi appears as an enormous, crimson lizard monster most of the time, but can also take the shape of a regular lizard. After he had arrived in Ao-tea-roa, Ngatoro's sister Kuiwai decided to follow him with her sister Haungaroa, as her husband had insulted her. Horomatangi spotted the two sisters before Ngatoro, who was resting. He traveled quickly through the waters to reach them. Some say this led to the large whirlpool three miles south of Motutaiko near Horomatangi Reef. He was blocked from the sisters, but used a plume of scalding steam to point toward Ngatoro, and the sisters knew where to point their canoe.

Some stories say that he then rested, and turned into a large black rock. Most say that he went on to dwell at Lake Taupo. Another story says that Horomatangi became acquainted with a fellow taniwha, Huru-kareao, who had taken to the residents of Rotoaira. Huru-kareao may even be Horomatangi's cousin.

Horomatangi was so impressed with them that he, too, vowed to protect them. A girl from this community went to the neighboring Arawa, but was mocked by them, and asked for revenge. The two taniwha rushed to avenge her, but unfortunately their massive waves caused greater damage to Rotoaira than to anywhere else. It was after this that Horomatangi took up his residence at Lake Taupo and began assaulting canoes there.

Horomatangi would often attack the Maori people who fished and sailed on the lake, lashing at the boats with water and storms and then pulling people down to the bottom of the lake. The blowhole and bubbling nature of the lake is attributed to the great power Horomatangi possesses. Even modern tales tell of Horomatangi overturning motorboats, even on cloudless days. Some say he is jealous of their power.

According to some stories, certain Maori chiefs with significant mana, or spiritual power, could cross the lake and repel the water dragon for periods of time to allow safe passage of boats. But if this was attempted by a chief with low mana, he would be swallowed up by the taniwha.

Strangely, Horomatangi also possesses a human familiar, Ati-a-muri, who does his bidding. Ati-a-muri, sometimes described as a Taniwha in the shape of a man, would lure travellers toward the lake, so that Horomatangi could take them. At twilight, Ati-a-muri would paddle a boat near the houses on the lake, trying to attract people close to the shore to their deaths. Also, seemingly associated with Horomatangi and perhaps even Taniwha themselves, are two dogs made of stone. These dogs inhabit the Karangahape cliffs on the western shore of Lake Taupo. When the night is foggy and dark, the dogs will howl and bark on the cliffs above the lake. It is taboo to disturb the dogs in any way, should they be found in stone form - lest one incur the wrath of Horomatangi.

Illustration:

Horomatangi here takes elements from the order Plesiosauria, a group of aquatic reptiles whose remains are found all over New Zealand

SOURCES:

Bane, Theresa. (2015.) Encyclopedia of Beasts and Monsters in Myth, Legend, and Folklore. Jefferson, North Carolina, USA: McFarland & Company, Inc.

Evans, Jeff. (2009.) Nga Waka O Nehera: The First Voyaging Canoes. New Zealand: Libro International.

Hayter, Maud Goodenough. (1957, 2000.) Folklore and Fairy Tales of the Canterbury Maoris. Christchurch, New Zealand: Cadsonbury Publications.

Keane, Basil. 'Taniwha', Te Ara. The Encyclopedia of New Zealand. Retrieved from http://www.TeAra.govt.nz/en/taniwha.

Maori Culture. 100% Pure New Zealand. Retrieved from http://www.newzealand.com/my/maori-culture.

O'Callaghan, Regan. (2011, June 9.) Horomatangi. Regan O'Callaghan. Retrieved from http://www.reganocallaghan. com/?p=123ori/Maori-Myths-Legend.

Reed, A. W. (1963.) Treasury of Maori Folklore. Hong Kong: Dai Nippon Printing Co.

Reed, A. W. (1964.) Maori Fables and Legendary Tales. Great Britain: C. Tinling & Co. Ltd.

Reed, A. W., and Calman, Ross. (2008). Taniwha, Giants, and Supernatural Creatures: He Taniwha, He Tipua, He Patupaiarehe. North Shore, New Zealand: Raupo.

Cultural Questions:

1. Do you see any connections between the myths from Hawaii and the myths from New Zealand?
2. What is the role of the Rainbow Serpent in many of the Australian Aboriginal myths?
3. What is the significance of having a creator being who is also responsible for punishment?
4. Compare the different Rainbow Serpent myths for similarities and differences.
5. Discuss the familial contexts of the Mo'o myths.
6. Bathala is a major deity in southeast Asia, discuss his role in the mythology of the area.
7. Many of these cultural communities have been impacted by colonization. How is this reflected in the myths?

CHAPTER 7: SOUTH AMERICA

Background:

Many legends appear in the documents of the colonizers about dragons/serpents/monsters in South America. The stories are elusive. Many Indigenous South American people simply do not consider it appropriate to discuss these creatures. Translation from Indigenous languages to Spanish to English often twists meanings. While archaeology documents the Chinchorro culture, 900-3500 before present, a culture where everyone of every rank was mummified, it only shows images of serpents/dragons on temples as far back as the Chavin culture, the oldest known Peruvian "civilization," dating to 4900 before present.

Lake Titicaca in southern Peru, a center for many South American cultural systems, including the Inca, is thought by some to be protected by a very large snake. What follows is a summation of the more persistent legends.

Amaru

Background:

Many snake/serpent deities exist throughout the Americas. Some sources state that the name of the Americas does not come from Amerigo Vespucci, the explorer and cartographer generally attributed with demonstrating to Europeans that South America was, in fact, a separate land mass. Instead, other sources claim that the name "America" is derived from the name, "Amaruca", translated means "Land of the Serpent". It is claimed that in the distant past, all people in the Western Hemisphere believed in the deity, Amaru Meru, who symbolized mystical wisdom and spiritual power.

The Tiwanaku civilization predates the Inca and some say that it is as old as 3500 before present although most date it to 300-1150 A.D./C.E. The myths most often relate to the Tiwanaku and Inca civilizations. The Tiwanaku were based in the Lake Titicaca basin where there was predictable and abundant rainfall. They practiced flooded raised field agriculture. Over time, the canals between the fields became fish farms.

Amaru Mach'ay is a cave just north of Cuzco engraved with images of elephants, condors, pumas, snakes and a llama. It is said to be the "portal cave of serpents" and considerably older than the Incan Empire. The carved rock is andecite which is full of quartz, an element often held as a conduit of spiritual energy and heat.

Incas often requested anacondas, closely associated with Amaru, as tribute. In addition to being a serpent deity, Amaru is also a name given to a community of Indigenous people.

Myth:

Amaru was a double headed winged serpent who lives underground, usually in the streams and springs throughout the mountains, especially near Cuzco. He is sometimes depicted as stretching across the sky, emerging from one pool and diving into another, making him seem quite similar to many of the Australian Rainbow Serpents. Because he is two-headed, he may either be drawn as a winged snake with two heads close to each other, or a snake with a head at each end of its body. If one looks to the second type of Amaru depiction, in hopping from spring to spring, one head would then be in one spring and the other head in a different spring. Amaru also incorporates some feline characteristics, and may have the heads of a serpent, bird, or large cat. Some call this culture bearer, Amaru Meru. It is said that Amaru once united all of the Americas in a single spiritual belief system. Many still believe that the Americas will once more be united under this powerful serpent (possibly female).

The chronicler Agustin de Zorate wrote down a story of a time when snakes overwhelmed the human population due to a great flood. The humans killed the snakes. Serpents were carved in many caves throughout South America by Chavin people, as well as others. Amaru in particular is seen along with a step pyramid motif.

Amaru is seen as the mediator between wamanis (spirits of mountains, sky, livestock and man) and the Pachamamas (earth, domesticated plants and women). The stories say that Amaru collects ceremonial offerings from the mountains/men and brings them to the fields/women, thus connecting these two through water, replenishing life. As a revolutionary force that destroys unbalanced systems and brings about order, Amaru has also been a powerful symbol of rebellions across time. In some areas, mach'aqway, the great serpent, may refer to Amaru. Offerings are still made to Pachamamas when creating terra preta in the Andes.

Illustration:

Amaru the mediator figure is described as a perplexing amalgamation of animals. For this drawing, the following Neotropical organisms were combined: rainbow boa (*Epicrates cenchria*), puma (*Puma concolor*), arapaima (*Arapaima leptosoma*), llama (*Lama glama*), and hawk (Accipitridae).

SOURCES:

Adrián Ambía, Abel. (1970.) Amaru: Mito i Realidad del Hombre. Lima, Peru: Pukara.

Amaru. Encyclopaedia Britannica. Retrieved from https://www.britannica.com/topic/Amaru.

Bolin, Inge. (1998.) Rituals of Respect: The Secret of Survival in the High Peruvian Andes. Austin, TV: University of Texas Press.

Edrick, Vann. Amaruca and America. Controversial Files. Retrieved from controversialfiles.blogspot.com/2013/01/amaruca-and-america.html.

Hidden Inca Tours: Explore Mankind's Hidden History. (2017) Retrieved from http://hiddenincatours.com.

Inca Trail Treks. Kondor Path Tours. Retrieved from www.kondorpathtours.com.

O'Neill, Patt. (2007, June 13.) Glossary of Terminology of the Shamanic & Ceremonial Traditions of the Inca Medicine Lineage. Inca Glossary. Retrieved from http://www.incaglossary.org/a.html.

Smith, S. (2011). Generative landscapes: the step mountain motif in Tiwanaku iconography. Ancient America, 12, 1–69.

Steele, Paul Richard and Catherine J. Allen. (2004.) Handbook of Inca Mythology. Santa Barbara, CA: ABC-CLIO.

The Moon Temple or Amaru Machay. Ancient Mysteries Explained. Retrieved from http://www.ancient-mysteries- explained.com/moontemple.html.

Iemisch

Background:

Myths come from the Tehuelche people and their culture in the pampas region of Patagonia, Chile and Argentina, where they have lived for thousands of years. The Tehuelche continue to tell their stories today.

Prehistoric caves in the region have images of felines, rheas, guanacos, left hands and geometric designs. Prehistoric cave drawings found in Mylodon Cave, for instance, seem to suggest a creature that had a mix of feline and aquatic characteristics. After these paintings were discovered, explorer Florentino Ameghino began collecting stories of the Iemisch with his brother, Carlos. Supposedly, they even collected pieces of Iemisch hide. The accounts of the Iemisch were attacked by fellow naturalists. However, French naturalist André Tournouer with an Indigenous guide came across an animal the guide called "Hymché" which seemed to match descriptions of the Iemisch, perhaps confirming at least a widespread belief in the creature.

At least one paleontologist claimed that the Iemisch is a Neomylodon. Could it be?

Myth:

The Iemisch is a water tiger, part serpent, and possibly part fox, with a hide covered in pointy brown hairs, who lives in the rivers of Patagonia. Some call it the "fox-viper" and others call it "water-tiger."

The stories say that the lemisch is a water monster who rises up from the rivers and grabs unsuspecting victims with its claws, or its long tail, then crushes them like a boa constrictor as it slowly sinks into the water. Its long serpent tail is said to be prehensile. Some accounts said that it was nocturnal and amphibious, equally at home in the water and on land, with webbed paws to give it an aquatic advantage. This fits with the Tehuelche concept that the night gives birth to evil spirits and monsters.

One particular myth is told of a Tehuelche Indigenous man who had a hide from an lemisch that had tiny bones protruding from it. One Indigenous boy told a story of meeting an lemisch on the road, but that the animal could swim in the river as well as walk on land. Another story is told of an lemisch who came down from the Andes on the rivers and landed on the shore of Rio Santa Cruz. Local residents fled in fear of their lives and the shoreline still bears the name of the incident, lemisch Aiken.

Illustration:

lemisch hails from the southern portion of the continent. Here, the fossils of *Thalassocnus*, a giant (2 meters) aquatic sloth, can be found. The large claws, coarse fur, and a prehensile tail are all traits of this strange animal, used here as inspiration for lemisch. The patterning of the red-tail boa (*Boa constrictor*) is used for the tail. †*Megatherium* is the genus name given to giant ground sloths which roamed South America during the Pliocene and existed at the same time as ancient humans. These animals were some of the largest mammals to ever walk the earth, sometimes weighing up to 4 tons and reaching 20 feet in length. While herbivorous, these animals did possess large claws on the hands and feet and did have a sturdy tail (although not quite as exaggerated as that described in the mythology of the lemisch). Megatherium skeletons have been uncovered all throughout Chile and Argentina, where the lemisch myth is prevalent. The myths are often told in the summer when the Ihuaivulu constellation appears on the northern horizon of the skies above Chile and Argentina.

SOURCES:

Adem, Teferi Abate. (2009.) Culture Summary : Tehuelche. New Haven, Conn.: HRAF.

Chiti, Jorge Fernández. (1997.) Cerámica indígena arqueológica argentina 2, 64.

Dekirk, Ash. (2006.) Dragonlore: From the Archives of the Grey School of Wizardry. New Jersey: Career Press, Inc.

Emperaire, J. and Laming, A. (1954.) La Grotte Du Mylodon (Patagonie Occidentale). Journal de la Société des américanistes, NOUVELLE SÉRIE 43, 173- 205.

Leandro M. Pérez, Néstor Toledo, Sergio F. Vizcaíno, M. Susana Bargo. (2018). Los restos tegumentariosde perezosos terrestres (Xenarthra, Folivora) de Última Esperanza (Chile). Cronología de los reportes, origen y ubicación actual. Publicación Electrónica de la Asociación Paleontológica Argentina 18(1), 1–21.

Heuvelmans, Bernard. (1958, 1995.) On the Track of Unknown Animals. New York: Routledge.

McCormick, Kylie. (2013, September 9.) Iemisch. Dragons of Fame. Retrieved from www.blackdrago.com/fame/iemisch.htm.

Megatherium. Prehistoric Wildlife. Retrieved from http://www.prehistoric-wildlife.com/species/m/megatherium.html.

Meurger, Michel and Claude Gagnon. (1998.) Lake Monster Traditions: A Cross-Cultural Analysis. London: Fortean Times.

Rose, Carol. (2000.) Giants, Monsters, and Dragons: An Encyclopedia of Folklore, Legend, and Myth. New York, USA: W. W. Norton & Company, Inc.

Wilbert, Johannes and Simoneau, Karin, Eds. (1984.) Folk Literature of the Tehuelche Indians. Los Angeles, California: UCLA Latin American Center Publications.

Whittall, Austin. (2012.) Monsters of Patagonia. Ushuaia: Zagier & Urruty Publications.

Ihuaivulu

Background:

Volcanoes have in the past and continue to erupt along the Pacific Rim. The Araucanian/Mapuche culture traditionally existed in this area of volcanic activity, Chile and Argentina. It was a peaceful farming culture until it was overrun by the Inca and later the Spaniards. Mapuche lived in extended family groupings, creating larger groupings for war.

They were well known for the textiles woven by the women. In spite of contact with other groups, many Mapuche communities were still largely independent until the late 19th century.

Myth:

In the area of coastal and southern South America where Chile and Argentina exist today, myths are still told in the Mapuche culture of Ihuaivulu, the seven headed dragon who lives in volcanoes and breathes fire. "Angry spirits" are said to inhabit volcanoes, put there by Ngen, a powerful spirit who ruled the air. Ngen imprisoned some bad spirits who challenged his rule. Occasionally some are released as serpents. They contribute to the volcanic eruptions. Some stories describe Ihuiavulu as "long, slinky and a serpentine body, covered with burnished copper and red colored scales" and he can most certainly breathe fire (Bane).

Illustration:

While not seven-headed like the Ihuaivulu myth describes, the lava lizards (*Microlophus sp.*) of the Andes mountains can be found resting on the volcanic rocks found there. Some can have bright orange coloration on the throat. Here, Ihuaivulu is drawn using elements from these lizards.

SOURCES:

Aguirre, Sonia Montecino. (2012.) Chile Precolombino. Museo Chileno De Arte Precolombino.

Andre-Driussi, Michael. (1994, 2008.) Lexicon Urthus, 2nd Ed. Albany, CA: Sirius Fiction.

Barber, Richard and Riches, Anne. (2000.) A Dictionary of Fabulous Beasts. Suffolk, UK: Boydell Press.

Dekirk, Ash. (2006.) Dragonlore: From the Archives of the Grey School of Wizardry. New Jersey: Career Press, Inc.

Dragons in Mythology, Legend & History, Part 2. (2014, July 6.) Myth Beliefs. Retrieved from http://www. mythbeliefs. info/2014/07/dragons-in-mythology-legend-history.html.

Mapuche. Minority Rights. Retrieved from https://minorityrights.org/minorities/mapuche-2/.

McCormick, Kylie. (2013, September 9.) Ihuaivulu. Dragons of Fame. Retrieved from www.blackdrago.com/ fame/ihuaivulu.

Piccardi, Luigi and W.Bruce Masse. (2007.) Myth and Geology. London: Geological Society of London.

Rose, Carol. (2000.) Giants, Monsters, and Dragons: An Encyclopedia of Folklore, Legend, and Myth. New York, USA: W. W. Norton & Company, Inc.

Ilwanci

Background:

Northern Peru and eastern Ecuador is the Homeland of the Jivaro people, who call themselves as well as other Indigenous groups "Shuar." They are mostly known for their head hunting and the production of tsantsa, or shrunken heads, a Jivaro custom that was mistakenly generalized to all South American Indigenous groups. Due to sensationalism around shrunken heads, a great deal of misinformation exists around Jivaro people, though some ethnographies have sought to correct this. The Jivaro are also notable for having defeated invading Spanish forces, gaining their independence for several centuries, and prior to Spanish invasion they had successfully defeated an Inca attack led by emperor Huayna Capac. In spite of accounts of their "warlike" behavior, and pressure throughout the 1900s to accept Ecuadorian laws and culture (including the abduction of Jivaro children into boarding schools), they have established overwhelmingly peaceful relations with their nearby Indigenous and non-Indigenous neighbors since the mid-1800s and stopped headhunting amongst themselves in the early 1900s. At that time, due to sales to European and American markets, the custom had already begun shifting from one based on revenge killings to an economic practice.

Horticulture is the main source of food for the Jivaro. It is supplemented when possible with hunting. Clearly defined gender roles relate to a belief that most things have either a male or female soul.

Myth:

Many myths exist in the Jivaro culture about a serpent who can change its shape, the Iwanci. An iwanci is actually one of three recognized types of souls that each human has. When it is in the body (or corpse, including shrunken heads), it is called a muisak. If no trophy is made from the body, the muisak escapes and is then known as the iwanci. The iwanci can take three shapes: the macanci, a poisonous snake, the pani, a giant boa constrictor or anaconda, or a tree that will fall and crush a victim. However, an iwanci can also appear to the murderer of its body in dreams, as a man or jaguar. Jivaro mythology is full of images of pairs of anacondas and jaguars. The anaconda is one of several animal spirits considered to be powerful beings in Jivaro spiritual beliefs.

Whatever shape the Iwanci takes, humans should beware. It enjoys devouring human beings - any it encounters!

Illustration:

Iwanci is shown here in his anaconda (genus: *Eunectes*) form. Anacondas are a group of snakes belonging to the boa family and are the largest living snakes by weight (growing over 500 pounds). Interestingly, fossils of the largest known snake (the now extinct †*Titanoboa*, hypothesized to reach over 40 feet in length) are known from Colombia.

SOURCES:

Iwanci. The Strange Myth World. Retrieved from https://thestrangemythworld.com/2015/05/20/iwanci/.

Bane, Theresa. (2015.) Encyclopedia of Beasts and Monsters in Myth, Legend, and Folklore. Jefferson, North Carolina, USA: McFarland & Company, Inc.

Harner, Michael J. (1973.) The Jivaro: People of the Sacred Waterfalls. Garden City, NY: Anchor Books.

Harner, Michael J. (1974.) The Sound of Rushing Water. In Native South Americans. Patricia J. Lyon, Ed. Boston: Little, Brown and Company.

McCormick, Kylie. (2013, September 9.) Iwanci. Dragons of Fame. Retrieved from www.blackdrago.com/fame/Iwanci.

Rose, Carol. (2000.) Giants, Monsters, and Dragons: An Encyclopedia of Folklore, Legend, and Myth. New York, USA: W. W. Norton & Company, Inc.

Lampalugua

Background:

The stories of Lampalugua also come from the Mapuche, who were called Araucanian by the Spanish. They reside in Chile. The Mapuche were one of three groups who were somewhat united as the Araucanian. The other two groups, the Pinuche and the Huilliche, were assimilated by the Inca and the Spaniards. It is the Mapuche that resisted vehemently and from whom we have stories of the Lampalugua.

Myth:

Few stories are known about Lampalugua. Could it be because this huge serpent with the gigantic claws devours all humans and cattle it comes in contact with? Some say it has coppery red scales, some say it is an enormous dragon or "fox-snake", but, really, little is known for sure about this predatory monster.

Some say that it is lives below ground in caves in the water there and that witches kiss its tail before entering the cave for practicing ceremonies. In 1922, a story about the Lampalagua was published, describing the monster as a living dinosaur like the plesiosaur. In the story an explorer falls asleep with his head on what he thinks is a tree trunk, only to be awakened as the tree trunk moved like a serpent. A fight ensued. The Lampalagua was wounded and went down into Inca Lake, into a cave to hide and heal.

Illustration:

Another Chilean dragon, Lampalugua is described as both a fox and a snake. Here, patterns from the dwarf tegu (*Callopistes maculatus*) add the reptilian aspect to this dragon, and the mammalian elements are drawn from the Patagonian gray fox (*Lycalopex griseus*).

SOURCES:

Bane, Theresa. (2015.) Encyclopedia of Beasts and Monsters in Myth, Legend, and Folklore. Jefferson, North Carolina, USA: McFarland & Company, Inc.

Dekirk, Ash. (2006.) Dragonlore: From the Archives of the Grey School of Wizardry. New Jersey: Career Press, Inc.

Mapuche. Minority Rights. Retrieved from https://minorityrights.org/minorities/mapuche-2/.

Meurger, Michel and Claude Gagnon. (1998.) Lake Monster Traditions: A Cross-Cultural Analysis. London: Fortean Times.

Rose, Carol. (2000.) Giants, Monsters, and Dragons: An Encyclopedia of Folklore, Legend, and Myth. New York, USA: W. W. Norton & Company, Inc.

Whittall, Austin. (2012.) Monsters of Patagonia. Ushuaia: Zagier & Urruty Publications.

Guarani Dragons

Background:

The story of the creation of the Guarani people of Paraguay, Brazil, Argentina and Bolivia is still told within the Guarani culture. Today, the Guarani language is one of the official languages of Paraguay and is popular throughout Patagonia, with speakers in Argentina and Brazil as well.

Jesuits among the Guarani in the 1600s created "reductions," small towns to centralize the Guarani. Their religious and cultural system, in spite of strict European edicts and conflicting European accounts of their beliefs, contributed to a reputation as peaceful villages. They had an egalitarian system of redistribution, and have traditionally utilized hunting, fishing, gathering, and small-scale farming. The Tupi and Guarani had as a central theme of their religious belief a "place without illness," which was conflated with the Christian concept of Heaven although it originally differed from Heaven in that it could be reached within one's lifetime. Some Guarani myths also have close ties to pre-colonial Carib and Pemon myths.

A number of Guarani groups lived on the coast of San Paolo and Santa Catarina relatively undisturbed until development boomed in the 1970s. Through the 1980s they were able, with allies, to maintain some land bases though their traditional mobility, used to cooperate with other Guarani communities, was reduced.

Tupa was the god who created all, including humans. He is associated with the sun. He was helped by the moon goddess, Arasy. Arasy helped Tupa to come down from the heavens to a hill in Paraguay where he created all.

The first humans that Tupa created were Rupave, the father of the people, and Sypave, the mother of all people. Together they had 3 sons and many, many daughters. The sons were Tume Aranda, the wisest; Marangatu, a great leader of the Guarani and the father of Kerana; and Japeusa, a liar and hustler.

In Guarani mythology, eclipses were caused by an Eternal Bat who would gnaw the sun or moon, and this bat along with the Blue Jaguar would eventually destroy the world.

Myth:

Tao, known as the spirit of evil, wanted Marangatu's daughter, Kerana, for his own. For seven days and nights he fought Marangatu, known as the spirit of goodness. But Tao won and kidnapped Kerana and forced her to bear seven sons.

The moon goddess, Arasy, was so angry that she cursed the seven sons and that is where we come to the stories of three of those sons, monsters who are also referred to as dragons.

Teju Jagua

The first son with a serpent body is Teju Jagua. His serpent body has the head of a dog, some say seven dog heads. The dog's eyes can send out flames. Teju Jagua lives in caves, guarding whatever treasures he has accumulated. He also guards fruits. The story is told that Teju Jagua has shiny scales because he likes to roll around in the gold and precious stones in his treasure hoard. Despite his fearsome appearance and abilities, and his name which translates into "fierce lizard" or "monster lizard," he is generally considered quite genial in disposition.

Mboi Tu'l

Mboi Tu'l was the second son. He has the body of a serpent and the head of a parrot who has a forked tongue, scaly skin and feathers all over his head. Mbloi Tu'l's voice is so loud that it can be heard over very great distances, causing fear in the hearts of all who hear it. He is the god of waterways and aquatic creatures.

Monai

Monai is the third son cursed by Arasy. Monai is usually referred to as the lord of air and open fields. He also has the body of a serpent, but instead of the head of a parrot, Monai is a two horned serpent. His horns act as antennae. The horns can hypnotize any who come near, making it easy to gather his prey for consumption.

Monai is said to live in caves even though he is the lord of fields. He steals many things, being a fan of robbery, and hides his fortune in these caves. If a village was raided or great treasures disappeared, Monai was usually blamed.

All of the sons were defeated by the beautiful Porasy, who, according to legend, offered to marry Monai but wanted to meet all of his brothers first. Porasy waited in a cave while Monai gathered his brothers. When all were in the cave, Porasy tried to escape, but was caught, so she yelled to alert the villagers outside to burn the cave with Porasy and all of the cursed brothers inside, successfully ending the cursed children of Tao.

Illustration:

The sons of Tupa are drawn together here. Mboi Tu'I attempts to reconcile the shape of a snake and that of the hyacinth macaw (*Anodorhynchus hyacinthinus*), Monai is an exaggeration of the strange and wonderful mata-mata turtle (*Chelus fimbriata*), and Teju Jagua is drawn with the head of a hellish dog-like animal.

SOURCES:

Adelaar, W. F. H. (2006.) Guaraní. Encyclopedia of Language and Linguistics. Oxford: Elsevier.

Holloway, April. (2015, April 22.) The Gods of Creation and Legendary Beasts of the Guarani. Ancient Origins: Reconstructing the Story of Humanity's Past. Retrieved from http://www.ancient-origins.net/myths-legends- americas/gods-creation-and-legendary-beasts-guarani-002937.

Métraux, Alfred. (1948, 1998). The Guarani. New Haven, Conn.: HRAF.

Narciso Rosicrán Colman. (1929.) Ñande Ypy Kuéra ("Nuestros Antepasados"). BibliotecaVirtual del Paraguay. Retrieved from https://web.archive.org/web/20070930224211/http://www.bvp.org.py/biblio_htm/colman/indice.htm.

Native Languages of the Americas: Guarani Indian Legends, Stories, and Myths. (2015.) Native Languages. Retrieved from http://www.native-languages.org/guarani-legends.htm.

Paraguayan Myths. (2006.) Project Paraguay. Retrieved from http://www.projectparaguay.com/myths.htm.

Pereira, Vicente Cretton. (2016.) Our Father, Our Owner: Master Relations between the Mbya Guaraní. Mana 22(3), 737-764.

Sá, Lúcia. (2004) Rain Forest Literatures: Amazonian Texts and Latin American Culture. Minneapolis, Minnesota: University of Minnesota Press.

Cultural questions:

1. Evaluate the theory that the word America comes from the myth of Amaru. Is there any evidence for this theory?

2. The Mapuche are an agricultural people who have managed to persist in colonized South America. Compare the dragon myths of the Mapuche. What are the similarities and differences?

3. The Guarani are a highly stratified cultural system. The dragon myths tell of familial conflict, yet again, in a stratified system. What do you theorize about this?

4. Discuss the prominence of serpents in the South American myths.

CHAPTER 8: NORTH AMERICA

Background:

The range of dragon myths in North America reflects the great range of cultural diversity on the continent. From a fire breathing dragon of the Seneca in the Lake Ontario region, to the Cherokee tales of the Uktena with their crystals, to the Zuni myth of Kolowissi, dragon myths abound throughout the many cultural systems. The following are just a sample of these culturally significant myths.

Az-I-Wu-Gum-Ki-Mukh'Ti

Background:

Az-I-Wu-Gum-Ki-Mukh'Ti was first described in print by E. W. Nelson, a European naturalist who was exploring Greenland.

Myth:

Among Greenland Inuit peoples, Az-I-Wu-Gum-Ki-Mukh'Ti was well known, and was said to look like a walrus but have the head of a dog, fangs, four legs like a dog, shiny black scales, and the tail of a giant fish. One blow from its powerful tail could kill a man, sink a canoe, and possibly even destroy a large ship that entered its territory. Some legends insisted that it guarded regular walrus groups, perhaps acting as a protector for them. Nelson nicknamed it the "walrus dog," perhaps because the Inuit found it inauspicious to say its real name.

Illustration:

This illustration of Az-I-Wu-Gum-Ki-Mukh'Ti was based off the recently-extinct Steller's sea cow (†*Hydrodamalis gigas*), a manatee-like animal which grew to 30 feet, the walrus (*Odobenus rosmarus*), and exaggerated features like the paw-like flippers which liken it to a dog.

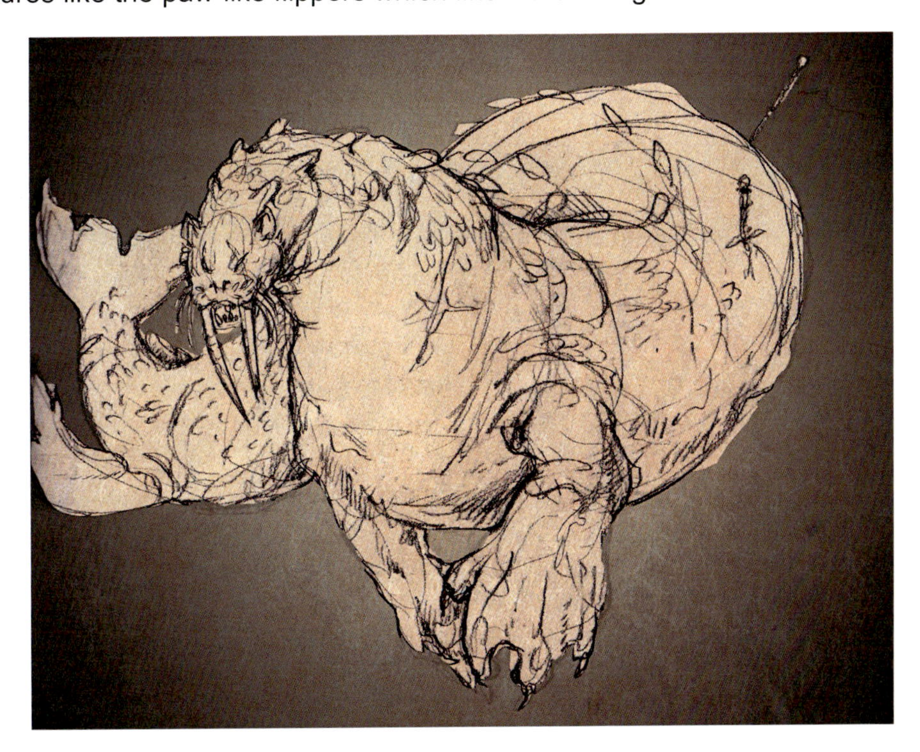

SOURCES:

Nelson, E. W. (1900) The Eskimo about Bering Strait. Extract from the Eighteenth Annual Report of the Bureau of American Ethnology. Government Printing Office, Washington.

Rose, Carol. (2000.) Giants, Monsters, and Dragons: An Encyclopedia of Folklore, Legend, and Myth. New York, USA: W. W. Norton & Company, Inc.

Mishi-ginebig

Background:

Many cultural systems have stories of floods that cover the earth. In North America, several of these myths share the story that the earth is created or re-created on the back of a turtle. Depending on the myth, a particular animal often dives deep into the water to find earth that is brought up and placed on the turtle's back. Therefore, many Indigenous groups in North America refer to the continent as Turtle Island.

While the hero of this tale, Nanabozho, is often described with masculine pronouns, he can also be seen as a genderless or fluid being. As Anishinaabe artist Wanda Nanibush writes, Nanabozho is "a transformer, a shapeshifter, not a he or she. Nanabozho doesn't believe in binaries."

Myth:

Mishi-ginebig is a creature whose name varies according to tribal community and dialect - Misi-kinepikw in Cree, Misi-ginebig in Oji-Cree, Mishi-ginebig in Ojibwe, Maji-skok in Abenaki, Msi-kinepikwa in Shawnee - but all translate to mean a great snake or, some say, a great evil snake. He is responsible for an earth-covering flood.

One Ojibwe legend involves a cultural hero, Nanabozho, who is often said to take the form of a rabbit. One day Nanabozho saw that a relative of his was missing, and noticed tracks like those of an enormous snake leading away from their home. The tracks led to a massive lake, and Nanabozho stood peering in. He saw the home of Mishi-ginebig, all his evil spirit friends, and in the center there was Mishi-ginebig himself. The snake's head was bright red, but the scales on his body shone like a rainbow in every color. However, all Nanabozho could see was that the snake was coiled around his cousin, whose hair floated around his still face in the water. Nanabozho meant to avenge his cousin's death, and so he began to rearrange the natural world. He sent away the clouds, calmed the wind, and then urged the sun to shine brighter, brighter, still brighter. Soon the lake began to steam, and then boil. Using his powers, Nanabozho gave himself the body of a tree stump and waited in hiding for his enemy to appear, and though many snakes fled the water, Mishi-ginebig stayed hidden in the darkest depths. As the water calmed, the snakes returned to it. Once again Nanabozho asked the sun to heat the waters even more, so that this time even Mishi-ginebig emerged on the lake with all of his snake friends. They covered the lake shore writhing and hissing in anger. As the day wore on, the snakes found shade and grew comfortable. All this time Nanabozho waited. Nanabozho waited even when the snakes whipped their tails around him, even when one clever snake smelled his scent and tried to throw the tree stump into the water. Finally, after night had fallen, he shed his disguise, carefully took out one of his arrows, and shot it into the heart of the great serpent. Mishi-ginebig's scream was so loud it woke all of the sleeping animals in the forests nearby. With his followers, he fled to his lake home where in a fit of rage they tore apart the body of Nanabozho's cousin.

Mishi-ginebig still had life left in him and was determined to get revenge. His snake companions helped him stir the waters of the lake until it began to rise and rise. Nanabozho ran from village to village warning everyone to go to the highest peak they could find. Eventually, even the mountaintops were overcome by the flooding, and Nanabozho helped create a large wooden raft to save as many people as he could. After days on the raft with no land in sight, the peaks of mountains finally began to emerge, and then grow ever larger as the water slowly went down. Returning to their homes, the people learned that Mishi-ginebig had at last died from his wounds, and his snakes returned to the lake fearing Nanabozho's deadly arrows.

Another version of the story provides an even happier ending. In this version, Nanabozho is the trickster rabbit and his relative is Wolf, who he warns never to jump over water - but Wolf defies Nanabozho's warning. Thus, he is captured and killed by Mishi-ginebig and his serpentine kin. Nanabozho turns himself into a tree stump, which the snakes are suspicious of, but eventually they decide that the stump cannot possibly be him. He then transforms and shoots the leader of the snakes with an arrow, but the great serpent does not die. Instead, Mishi-ginebig sends for Frog to tend to his wounds. Nanabozho intercepts Frog's mission, kills her, and wears her skin as a disguise. When he gets very close to Mishi-ginebig, he shoves the arrow even deeper into his heart and kills him. As the serpent's relatives panic, he steals back the fur of Wolf. Floods cover the land following Mishi-ginebig's demise, but Nanabozho and the other animals find safety on a wooden raft. There, Nanabozho breathes onto Wolf's skin and brings his beloved relative back to life. Then, the animals take turns diving for earth. Muskrat succeeds in finding some earth, and Nanabozho uses it to recreate the world anew.

Illustration:

While there are no obvious red-headed snakes in the northern United States or Canada to represent Mishi-ginebig, the northern ring-neck snake (*Diadophis punctatus*) comes closest. When frightened, this non-venomous serpent will flip its lover half over to reveal its red/yellow underside and curl its tail in a spiral. This almost looks like a read head when viewed from afar.

SOURCES:

Boutet, Michel-Gerald. (2015.) The Great Long Tailed Serpent: An iconographical study of the serpent in Middle Woodland Algonquian culture. Midwestern Epigraphic Society. Retrieved from http://www.midwesternepigraphic.org/ The%20Great%20Long%20Tailed%20Serpent.pdf.

Edmonds, Margot and Ella Clark. (1989.) Voices of the Winds: Native American Legends. Castle Books.

Kawbawgam, Charles, Kawbawgam, Charlotte, LePique, Jacques, Kidder, Homer H., and Bourgeois, Arthur P. (1994.) Ojibwa narratives of Charles and Charlotte Kawbawgam and Jacques LePique, 1893-1895. Detroit : Wayne State University Press.

McLeish, Kenneth. (1996). Myth: Myths and Legends of the World Explored. New York, NY: Faces On File, Inc.

Messer, Ron. (1989.) A Structuralist's View of an Indian Creation Myth. Anthropologica 31(2), 195-235.

Nanibush, Wanda. (2018.) Nanabozho's Sisters. Canadian Art 35(3), 37-37.

Willoughby, Charles C. (1935.) Michabo the Great Hare: A Patron of the Hopewell Mound Settlement. American Anthropologist 37(2), 280-286.

Gaasyendietha

Background:

The Lake Ontario region of North America is known to be a place with a plethora of meteor showers, especially in the summer. Within Lake Ontario itself is the Charity Shoal crater, thought to be an impact crater from more than 400 million years ago. This myth comes from the Seneca, a nation in the Haudenosaunee Confederacy, also known as the Iroquois. The Haudenosaunee were five feuding tribal nations who formed an alliance and created one of the oldest democracies in the world, a government system which still exists today. While most farming societies are patrilineal and patrilocal, the Haudenosaunee were both skilled farmers and consciously organized themselves as a matrilineal, matrilocal cultural system. Their cultural and governmental ties have maintained a peaceful democracy for nearly a thousand years.

Myth:

Gaasyendietha is perhaps the only fire-breathing dragon of North American legend. According to Seneca tales, Gaasyendietha lives in Lake Ontario but visits other ponds and lakes. It also flies across the sky on a trail of fire and was said to have come to earth on a falling star, earning it the title of "meteor dragon."

Sightings of Gaasyendietha have taken place throughout history; a few cases are in the writings of explorer Jacques Cartier, in a ship's report in 1817, and the story of two children who claimed to have seen the dragon in 1829. Sightings have occurred as recently as the 1970s, making this ancient legend thoroughly modern.

Illustration:

Gaasyendietha here takes a form inspired by the lake sturgeon (*Acipenser fulvescens*). Lake sturgeon, found (among other freshwater systems) in the Great Lakes, are covered in bony scutes and can grow over 7 feet long.

SOURCES:

Dekirk, Ash. (2006.) Dragonlore: From the Archives of the Grey School of Wizardry. New Jersey: Career Press, Inc.

Moffat, Charles Alexander. Gaasyendietha. Sea Serpents of Canada. Retrieved from http://www.lilith-ezine.com/ articles/2005/canadian_seaserpents.html.

Rose, Carol. (2000.) Giants, Monsters, and Dragons: An Encyclopedia of Folklore, Legend, and Myth. New York, USA: W. W. Norton & Company, Inc.

Uktena

Background:

The Cherokee are the southernmost cultural system with an Iroquoian language family. Like the Haudenosaunee (Iroquois), the Cherokee are matrilineal and agricultural, focusing their farming efforts on corn, beans and squash. Cherokee comes from a Creek word meaning people of a different speech. They were considered one of the "five civilized tribes" by Europeans, and skillfully incorporated European elements into their society while maintaining Cherokee traditions. Today, the Cherokee in Oklahoma are one of the largest Indigenous nations in North America, and the Eastern Band of Cherokee that remained in the southeast, evading capture and expulsion on the Trail of Tears, also remain. Stories of horned serpents exist in several cultural systems in North America. Uktena is one of the best known, and is typically depicted as a large serpent with a rack of antlers, as are many mythological snakes across Indigenous southeastern groups. Many depictions, such as a modern portrayal by Roy Boney Jr., also show feathered wings on the uktena. A much older Mississippian Moundville pot similarly shows a large, antlered snake with wings.

Myth:

Long ago, the homelands of the Cherokee were inhabited by great and powerful monsters - insects large enough to carry off children, giant frogs and lizards, and of course, monstrous serpents. Some of these serpents walked the earth with legs and were so large their heads cleared the tops of trees, like the ustutli snake who would cry like a young deer to lure in hunters. Yet the greatest of all the serpents was the Uktena, and he was once a man.

As it happened, the Sun had grown angry with the people of earth for making faces when they looked at her, not understanding that her rays hurt their eyes and made them twist their features. Her brother, the Moon, begged her not to take revenge on the humans, since he found them all very lovely illuminated by his soft light. But the Sun could not be dissuaded. Every day, she rose in the sky, stopped briefly for lunch at her daughter's home which lay between the sun and the earth, and then shone her rays so hard that the people became very sick with the heat. Humans asked for help from the Little People who lived among them, rarely seen but often kind and helpful. The Little People agreed to work their magic and help. They changed two young men into snakes, one into a spreading adder and the other into a copperhead. The snakes waited outside the Sun's daughter's home, but being blinded by the Sun's powerful rays all the spreading adder could do was spit his venom wildly, and the copperhead turned and ran. So the Little People tried again, turning two men into a rattlesnake and an Uktena. The rattlesnake was small but quick, with incredible venom. Uktena was massive in size with mighty horns and many magical abilities which came from the crystal on its head.

So these two champions set out for the home of the Sun's daughter to wait and attack the Sun. At the first movement the rattlesnake saw, he struck - but his fangs landed on the Sun's daughter and not the Sun herself. Grief-stricken, the Sun sent humans to retrieve her daughter from the land of ghosts, but the mission failed and she wept so much that the land was flooded for a long time. Uktena, enraged that the rattlesnake had missed his mark - but also that he himself had failed the mission against the Sun - became twisted and bitter until he could no longer live near humans. He was banished deep in the mountains, though he left behind many gruesome uktena snakes like himself. Taking after the first of the uktena, their scales shone like rainbows, except for the great crystal on their heads called an ulunsuti, which is transparent with a blood-red streak running through it. They could kill with their pestilent breath, and seeing an uktena usually meant death or great calamity even if one survived the encounter - and most did not. Due to their beauty, most people who saw the uktena became mesmerized and ran toward its expectant jaws.

A great medicine man named Aganunitsi was the only man to ever kill one of these massive serpents. He had been fighting against the Cherokee, but was finally captured by them. The Cherokee were unsure what to do with him, and were considering killing him due to his dangerous magical abilities, when he offered to do any task they wanted should they spare his life. No one believed that he could kill an uktena, but he assured them that even this could be done with his great skills. So Aganunitsi set off into the mountains, where all the dangerous creatures lived. First he saw a blacksnake of incredible size, but he was not frightened. Then he came upon a moccasin snake, also larger than any that had ever been seen before, but he pressed on. Next, an enormous green snake appeared in his path. At this point many of the people who had accompanied him turned back, but he found a way around it and kept going. Once he reached the Bald Mountain, he found a lizard larger than a man, but he was not distracted by it. Even further south, he found a frog whose mouth could swallow a child whole.

Many people had turned back now, but Aganunitsi pressed onward, laughing at those who could be frightened by a mere frog, regardless of its size. At the Gap of the Forked Antler there were monstrous reptiles roaming everywhere, but he quietly moved through it. Now he was alone, everyone else having abandoned the mission. He found a deep lake, and though it too was filled with great beasts - biting turtles and immense sunperches and dangerously large fish - uktena was not here either. He kept going south, and finally saw a gaping cave on the side of a mountain. Sneaking up to the cave, he saw the uktena sleeping. He knew he would have to shoot the seventh spot from the serpent's head, for that was where its heart lay, but first he made preparations for the battle. He set a fire around the base of the mountain, and dug a deep trench inside of the fire. Then he returned to the cave and shot the uktena in the heart. The monster woke with a roar and, spotting Aganunitsi, slithered after him immediately. Aganunitsi ran headlong down the mountain and jumped over the trench and through the fire. The uktena's wrath was great, and he spit venom everywhere, but it was evaporated in the fire Aganunitsi had built. Only one small drop reached Aganunitsi's head before he jumped through the flames. Toxic blood gushed from the uktena's wound, but it collected in the trench and did not reach Aganunitsi. Finally the uktena collapsed and rolled down the mountainside, dead. Aganunitsi called upon the birds, who are the snake's natural enemies, to come feast on the monster. He left them to their task for seven days, and when he returned all that was left of the snake was the crystal on its head, the ulunsuti.

Aganunitsi took the stone and made his journey back to the village. Although he did not realize it, a small snake had grown into his hair at the very spot where the uktena's venom had landed on his head. He made quite a sight walking into the village, but his life was charmed from that point on, because the ulunsuti crystal gave him great power. It granted him luck in hunting, in romance, in controlling the weather, and it also allowed him to see into the future. No one knows what happened to the ulunsuti when Aganunitsi died after a long and prosperous life. Most say that it was buried with him, for the ulunsuti crystals are never meant to be looked upon by any person but their rightful owner, who must care for them and be buried with them or else they will cause great chaos for seven years before they are willing to accept their master's death and become quiet again.

Although humans would sometimes risk hunting the uktena for these powerful ulunsuti stones, the greatest enemy of the uktena were the tlanuhwa, or thunderbirds, and it was the thunderbirds who ultimately ended the days of the uktena. The Cherokee had always gotten along well with the thunderbirds, until one day these giant birds began picking up children to feed their young, who were about to hatch. The thunderbirds swept down on them, wrapped their talons around the children's arms, and carried them high onto a cliffside. The parents were stunned, but went after the thunderbirds immediately. One brave man volunteered to distract the thunderbirds while other warriors scattered the unhatched eggs. In the midst of these distractions, the children were led to safety. The uktena, in their underwater homes, heard the cries and emerged to see what was happening. They came up to the cliff where the humans were stealing back their children from the thunderbirds, and saw the falling eggs. The uktena waited there in the water, eating the eggs as they dropped. The thunderbirds saw that the humans had escaped with their children, and that the uktena had eaten all of their eggs, their future generations lost forever. In their anger, the thunderbirds flew at the uktena, lifting them into the sky and tearing them to shreds with their claws. They hunted down all the remaining uktena, and scattered their remains all over the mountains.

Thus ended the uktena forever, leaving as the only evidence of their existence the blazing clear ulunsuti crystals from their heads.

SOURCES:

Roy Boney, Jr. Artist. Retrieved from http://royboney.com/roy-boney-jr-art#/id/i3404041.

Chavez, Will. (2012, April 17.) Cherokees create artwork using guitars. Cherokee Phoenix. Retrieved from https://www.cherokeephoenix.org/Article/Index/6175.

Mooney, James. (1900.) History, Myths, and Sacred Formulas of the Cherokees. 19th Annual Report of Bureau of American Ethnology 1897-98, Part I.

Swann, Brian, Ed. (2004.) Voices from Four Directions: Contemporary Translations of the Native Literatures of North America (Native Literatures of the Americas). Lincoln: University of Nebraska Press.

Uktena-like monster incised on a Moundville pot. Retrieved from https://library-artstor-org.silk.library.umass.edu/asset/ARTSTOR_103_41822001453990.

Wolfe, David Michael Wolfe. Legend of the Tlanuhwa and the Uhktena. Cherokees of California. Retrieved from http:// www.powersource.com/cocinc/articles/tlanuhwa.htm.

Estakwnayv (or stvkwvnaya)

Background:

A mix of Muscogee Creek people and smaller communities Indigenous to Florida, Seminole developed an independent cultural system when they formed a nation. Their two traditional languages, Mikasuki and Creek, are both part of the Muskogean language family. Living in the swampy Florida environment, they became well acquainted with snakes and reptiles. Like many other southeastern Indigenous groups, they have a matrilineal clan systems with women maintaining positions of high status.

Myth:

Among the Seminole, there are many different legendary snakes of cultural significance. One legendary snake is the first member of the snake clan. A young hunter, he ventured into a forbidden part of the swamp and hunted fish there, knowing that the elders had warned him against it. Slowly his body began to transform into a snake's body, his legs fusing together and growing a layer of slick scales.

Another great snake is the "sharp-breasted snake," snakes which leave an imprint in the ground so you can see where they moved over the earth. The Mikasuki, some of whom joined the Seminole and some who are now known as the Miccosukee Tribe of Indians in Florida, traditionally described it as having large invisible wings. "The sharp-breasted snake is a serpent which goes along with its head up and its breast advanced. It is rarely seen but you can tell where it has passed along. These snakes are not thought to be very long, but they appear to vary in size. The largest would probably measure a foot and a half in diameter. With its sharp breast this snake tears up the earth, making a deep furrow. It can cut through the roots of trees, making the trees keel over, and throw mud high up on the trunks of trees nearby" (Grantham 34-35). A "sharp breast" certainly brings to mind the image of a dragon.

Yet another is the tie-snake, huge creatures capable of eating horses and cattle whole. Most people called these snakes estakwnayv, and believed that eating taboo foods might cause you to transform into one. One legend in particular states that a Muscogee Creek chief who refused to go to war with Tecumseh was about to be destroyed, along with his village, by the surrounding tribes who supported the war. Finally, he sent his son to provide a message to a nearby chief, but the gift the son was supposed to present was lost in a river when he foolishly tried to skip it across the water. However, when the boy swam out to get it, he was grabbed by several tie snakes and taken to their king, who vowed to stand by the boy's father when trouble came to him. True to his word, when a warring tribe was en route to attack the peaceful chief's village, the tie snakes wrapped around the warriors' wrists and ankles, tying them up on the ground. The tie snake may have horns, or may simply look like a massive snake.

Some stories say that the tie snake is the horned serpent, that they are one and the same, but others believe that they are two different snakes. Both lived in dark watery places and were immensely strong and powerful. But it is also said that the horned serpent has one large horn, which could be ground into powder and used as an aphrodisiac or as a benefit to the hunting skill of those who kept the horned serpent calm long enough to take it from him. And while the tie snake is usually described as being black all over, the horned serpent Sint Holo is iridescent, gleaming in the light, with a crystal on its forehead which can grant powers.

One final story says that the horned serpent was also once a man in the early days of the world. His twin brother was claimed by the earth and given skills with plants. But the water claimed the other brother, who was to rule the lakes and rivers and seas. When he entered the water, he began to transform into a snake from the waist down. It is said that when his brother calls on him, he will appear to help, and that he may even reveal himself to very wise young men. Yet for most, he is a dangerous creature who one would be wise to avoid.

Illustrations:

The horned serpents of this chapter are drawn here together to better show their diversity. The sharp-breasted serpent is based off the American alligator (*Alligator mississippiensis*), because like the myth, alligators regularly dig up the land to create "gator holes," which are ecologically important. Estakwnayv, one of the tie-snakes, is shown here with moose (*Alces alces*) antlers and a water moccasin (*Agkistrodon piscivorus*) body. Uktena, another horned serpent, is drawn with the body of a eastern diamondback rattlesnake (*Crotalus adamanteus*) and the horn of a white-tailed deer (*Odocoileus virginianus*). A smoky quartz crystal is embedded in Uktena's forehead.

SOURCES:

Dekirk, Ash. (2006.) Dragonlore: From the Archives of the Grey School of Wizardry. New Jersey: Career Press, Inc.

Grantham, Bill. (2002.) Creation Myths and Legends of the Creek Indians. University Press of Florida.

Haney, Enoch Kelly. Legend of the Snake Clan. Seminole Nation Museum, Wewoka OK. Retrieved from https://www.seminolenationmuseum.org/m.blog/23/legend-of-the-snake-clan.

Hudson, Angela Pulley. (2010.) Creek Paths and Federal Roads: Indians, Settlers, and Slaves and the Making of the American South. North Carolina: University of North Carolina Press.

Methvin, Rev. J.J. (1927, December.) Legend of the Tie-Snakes. Chronicles of Oklahoma 5(4).

Mitchell, Deborah. King of the Waters: The Legend of the Horned Water Serpent. Southeastern Oklahoma State University. Retrieved from http://www.se.edu/nas/files/2013/03/2ndsymposiumpart5.pdf.

Schultz, Jack Maurice. (2008.) The Seminole Baptist Churches of Oklahoma: Maintaining a Traditional Community. Norman, OK: University of Oklahoma Press.

Kolowissi

Background:

The Zuni Nation live mostly in a pueblo in southwestern New Mexico. Their first contact with Spanish conquistadors was violent, with the conquistadors typically outlawing Indigenous religious practices, assaulting Indigenous women, and putting down resistance movements with brutal violence. These negative relations resulted in the Pueblo Revolt of 1680 when, along with other Southwestern Indigenous groups, they successfully expelled the Spanish for several decades. Traditional religious practices continue to be maintained, though some are considered private. The Zuni language is a linguistic isolate, different from all other languages, though it shares a few words with several surrounding Indigenous cultural communities. Even prior to the colonial period, the Zuni were able to farm corn in the arid environment of the southwest through a strong focus on water conservation. The Zuni maintain a matrilineal clan system and a strong ceremonial base.

Myth:

There are several stories about the great serpent Kolowissi - most good, some bad. Kolowissi gave the Zuni life, but for at least one girl, he also took life away.

One girl, long ago in Zuni history, did not heed the warnings to respect Kolowissi. She was a lovely young woman, but had become obsessive in her cleanliness, never allowing dirt to touch her skin for long. She became so fixated on remaining clean at all times that she built a room separate from her entire family, and went many times a day to the spring which everyone knew to be Kolowissi's sacred home.

Kolowissi allowed her transgressions for some time before seeking his retribution. One day, washing her arms in the spring, the young woman looked up to see a baby in the shallow water nearby. She scooped up the baby and took it to her private room, where her family eventually found her with her newfound child. They knew something was wrong about the sudden appearance of a child, saying to themselves that no mother would leave her baby in a spring. But the young woman would not be parted from the child, and so as night fell the woman held the baby close and fell asleep.

As the moon rose, the baby began to change. It grew long, and its arms and legs disappeared, melting into its sides. Scales covered its body. It finally grew so large in the small room that its tail was forced into its own mouth. The woman awoke, and the snake wrapped its coils around her and broke free from the house, dragging her back to his spring to be his bride forever.

However, despite Kolowissi punishing individuals for immoral behavior, not everyone had reason to fear Kolowissi. Years earlier, he used his massive body to hold floodwaters at bay while many Zuni ran up the side of a mountain named Dowa Yallane, or Corn Mountain. Once they were stranded on the mountain, they quickly ran short on food. Kolowissi heard their prayers and returned to the mountaintop, coming to them from the west. He opened his mouth and fresh waters poured out, followed by meat and seeds. Some ceremonies are still performed to honor this act, with seeds spilling from a carved Kolowissi's mouth. Some stories say that a young man and a young woman were volunteered as sacrifices, after which the waters pulled away from the mountain. But by returning to the Zuni, he saved the people from starvation and chose to live close to them, in the earth and its waters, so as to always be nearby.

It is said that to Tewa peoples Kolowissi is known as Avanyu, and to the Hopi as Palulukon or Palolokong. He may be related to Quetzalcoatl, since he is often pictured as a plumed serpent.

Illustration:

Kolowissi is shown here using elements from traditional plumed serpent dolls, as well as patterning of the southwestern Gila monster (*Heloderma suspectum*), a venomous lizard.

SOURCES:

Aftandilian, David. Ed. (2006.) What are the Animals to Us?: Approaches from Science, Religion, Folklore, Literature, and Art. Knoxville, TN: University of Tennessee Press.

Erdoes, Richard and Alfonso Ortiz. (1984.) American Indian Myths and Legends. New York, NY, USA: Pantheon Books.

Mills, Barbara J. and T.J.Ferguson. (2008.) Animate Objects: Shell Trumpets and Ritual Networks in the Greater Southwest. Journal of Archaeological Method and Theory 15, 342.

Parsons, Elsie Clews. (1923.) The Origin Myth of Zuni. The Journal of American Folklore 36(140), 135-162.

Palulukon

Background:

The Hopi occupy several Pueblo villages in the southwest of North America. Theirs is traditionally a matrilineal, matrilocal cultural system which is heavily focused on growing corn. Like the Zuni, they must carefully manage water stores in order to grow corn. The word for corn and the word for a woman of childbearing age is the same. Both women and corn are highly regarded in this cultural system. The term "Hopi" may also be used to describe their belief system, which emphasizes care for land, ethical treatment of others, and harmony with all things. Despite their strong cultural emphasis on maintaining peace, they did help drive out the Spanish in the Pueblo Revolt of 1680, after multiple cases of mistreatment by the Spanish. Although missionaries from Spain, and later the United States, attempted to convert the Hopi from their traditional religious practices to various denominations of Christianity, the traditional ceremonies of the Hopi are still widely practiced. Art forms such as weaving, pottery, and silversmithing are still maintained as well.

Myth:

In many societies, snakes - and dragons - are closely associated with water. It makes sense, then, that for a people like the Hopi who live in a desert environment great serpents are respected and honored more than they are feared. The great rivers of the southwest twist and turn like a snake's body, and earthquakes shake the ground like the rattlers of a rattlesnake.

Hopi stories describe the palulukon as great serpents who control the waters, including thunderstorms and the rain. All water is connected, one large ocean under the land which makes up a lower world beneath our own. The many palulukon (or palulukonti) in the world live in pools of water and, from there, watch over the lives of humans and animals alike. Though palulukon are generally helpful, they may also turn violent when they are mistreated or disrespected. Just as they maintain balance in nature, they also ensure balance in social relations. When this balance has been disturbed by humans, the palulukon may withhold water, dry up wells, create lightning storms, or even create earthquakes in their displeasure. In the past, at least one palulukon had destroyed Palatkwapi, the homeland of a group of misguided humans, by flooding it. The city had acted wrongly, and when the land was flooded they sacrificed a young man and a young woman to the serpent. However, the palulukon took them under the water, then into the earth, where he kept them safe and gave them lessons. When they returned to the earth's surface, the waters receded and the people moved to another place to rebuild in a better way.

The seemingly destructive actions of palulukon may also be taken to punish individuals for immoral behavior. However, not all natural disasters are caused by the anger of palulukon. Some stories say that the world floats on pools of water which are carried on the backs of two mighty palulukon. In these stories, earthquakes and disruptions in the water supply can be the result of the two dragons turning and twisting to shift their position out of pure exhaustion.

As symbols of water - and, by extension, fertility and life - palulukon are considered dangerous for pregnant women and children. Although there are female palulukon, in most stories they are described as male. There have also been reports of women becoming pregnant mysteriously at celebrations of palulukon, which take place in early spring when snakes are just beginning to emerge from the chill of winter.

Illustration:

Two palulukon are drawn here underground, supporting the land above with their backs.Their design is inspired by the charismatic western hognose snake (*Heterodon nasicus*), which spend a great deal of time underground. They are considered rear-fanged snakes and that element is exaggerated here.

SOURCES:

Dekirk, Ash. (2006.) Dragonlore: From the Archives of the Grey School of Wizardry. New Jersey: Career Press, Inc.

King, William R. (1987.) Dionysos Among the Mesas: The Water Serpent Puppet Play of the Hopi Indians. American Indian Culture and Research Journal 11(3), 17-49.

Mills, Barbara J. and T.J.Ferguson. (2008.) Animate Objects: Shell Trumpets and Ritual Networks in the Greater Southwest. Journal of Archaeological Method and Theory 15, 342.

Rose, Carol. (2000.) Giants, Monsters, and Dragons: An Encyclopedia of Folklore, Legend, and Myth. New York, USA: W. W. Norton & Company, Inc.

Turner, Patricia and Charles Russell Coulter. (2000.) Dictionary of Ancient Deities. New York, NY: Oxford University Press.

Von Grunebaum, G.E. and Roger Caillois. (1966.) The Dream and Human Societies. Berkeley, CA: University of California Press.

Quetzalcoatl

Background:

Quetzalcoatl is a god of the Aztec, Maya, and Toltec peoples whose name, from the Nahuatl language, means "feathered serpent." The Maya may be the oldest of the three cultural systems, and usually referred to Quetzalcoatl as Kulkulkan, but all three were closely connected. It's possible that the stories evolved as they passed on to each group, with each one interpreting the powers and disposition of Quetzalcoatl differently.

Among the Aztec, or Nahua, stories are varied as to how Quetzalcoatl came to be born. Some say God came to his virgin mother Chimalman in a dream, or that she became pregnant with him after swallowing an emerald. Another says that Mixcoatl was his mother, and that she became pregnant after being struck in the stomach with an arrow. Other stories have Coatlicue as his mother, and yet others say that his parents were Ometecuhtli and Omecihuatl, also known as Tonacateuctli and Tonacacihuatl, and that he and his three brothers guard each of the four cardinal directions. Quetzalcoatl the White guards the west and is the god of light, justice, mercy and wind; Huitzilopochtli the Blue guards the south and is the god of war; Xipe Totec the Red guards the east and is the god of gold, farming, and springtime; and Tezcatlipoca the Black guards the north and is the god of judgment, nighttime, sorcery, deceit, and the earth. There is debate on how much contact with Christianity has influenced the story of Quetzalcoat's birth, especially the versions depicting virgin conception.

Myth:

There are stories about Quetzalcoatl creating the earth with his opposite brother Tezcatlipoca, in order to make a balanced world. They created the earth and sky and saw that the earth was flat and featureless. They went into their great snake forms and chased after the great female monster Tlaltcuhtli, herself a reptile, tearing her into pieces over the land so that her hair and skin could become springs and her eyes and nose could become caves and her mouth could become valleys and mountains. Angry over the loss of her physical form, she demanded human sacrifice, which Tezcatlipoca consented to; yet Quetzalcoatl, in most stories, was less eager to perform. This was the first of many disagreements between Tezcatlipoca and Quetzalcoatl. The balance between the two brothers, so different from one another in so many ways, ultimately proved difficult to achieve and the ensuing fights between the two often destroyed the earth.

One of the reasons many people believe that Quetzalcoatl opposed his brother on the issue of human sacrifice is that in many stories, he has helped humankind. Known as the lord of the morning star or lord of the star of dawn, he is credited with inventing the calendar, books, and giving people the knowledge to grow corn. He also brought humanity back from the dead. After the fourth cycle of the world ended with so many natural disasters that all humans were wiped from the earth - possibly because of yet another epic battle between Quetzalcoatl and his brother Tezcatlipoca - Quetzalcoatl traveled to Mictlan, the underworld, to gather the bones of the exterminated human race. However, the ruling gods of the underworld, Mictlanteuctli and Mictlancihuatl, were unwilling to give up the dead. They told Quetzalcoatl that he could have the bones if he could blow on a conch shell that had no hole, and make it sound. Quetzalcoatl called upon worms to burrow holes in a sealed conch shell, and then he filled it with bees. Shaking it in front of Mictlanteuctli, it appeared to come to life and sound on its own, angering the underworld god greatly. To appease his anger, Quetzalcoatl told the underworld god that he had changed his mind about bringing humans back to life - but Quetzalcoatl had already gathered the bones and planned to smuggle them out. Mictlanteuctli did not believe Quetzalcoatl and, suspecting some kind of trickery, had a pit dug at the exit of the underworld. Quetzalcoat fell into the pit, mixing all the bones together so that the bones of men and women were indistinguishable. He escaped from the hole, but had to enlist the help of another snake deity, the goddess Cihuacoatl, to help sort them all out. He then cut himself on his tongue, his earlobes, his calves, and his groin, and used his own blood to bring humans back to life. Afterwards, he taught them many skills and arts, and was associated with learning and scholarship.

Illustration:

When the resplendent quetzal bird (*Pharomachrus mocinno*) flies, its long emerald tail sways in the wind like a giant flying snake. Quetzalcoatl is drawn here using elements from this bird, as well as those of raptorial dinosaur fossils found in this region. He wears a conch around his neck and also a mask (which were important to Aztec religious ceremony).

SOURCES:

Aldington, Richard and Delano Ames, Transl. (1959.) New Larousse Encyclopedia of Mythology. Hong Kong: Prometheus Press.

Bierlein, J.B. (1999.) Living Myths: How Myth Gives Meaning to Human Experience. New York, NY: Ballantine Books.

Carrasco, David. (1982.) Quetzalcoatl and the Irony of Empire: Myths and Prophecies in the Aztec Tradition. Chicago, IL: University of Chicago Press.

McLeish, Kenneth. (1996). Myth: Myths and Legends of the World Explored. New York, NY: Faces On File, Inc.

Quetzalcoatl. (2013, August 1.) Ancient History Encyclopedia. Retrieved from http://www.ancient.eu/Quetzalcoatl/.

Smith, Michael E. (2003.) The Aztecs 2nd Ed. Malden, MA: Blackwell Pub. Ltd.

Cultural Questions:

1. Though Az-I-Wu-Gum-Ki-Mukh'Ti attacks Inuit and non-Inuit alike, would sinking a large ship — which would be from a colonizing population — make him something of a protector of the Inuit as well?
2. Is it significant that Az-I-Wu-Gum-Ki-Mukh'Ti has features of many animals that Inuit traditionally hunt?
3. Many cultural groups believe that it is bad luck to call out the name of a monster. What does this say about human conceptions of language and the importance of speech?
4. What differences do you see between the version of the story where Nanabozho is a human man, and the one where he is a rabbit?
5. Some scholars and storytellers believe that the rabbit version of the story is the earlier version. Why did it change over time? Does this tell us anything about cultural changes within groups related to the Ojibwe?
6. What ecological phenomenon does the story explain?
7. What role does listening play in the story? Are there any lessons modern readers could pull from it?
8. Is it unusual for Gaasyendietha to have a home, but go visit other places? Are dragons typically associated with a limited geographic range?
9. What are the roles of meteors in Haudenosaunee stories and histories? What do they signify?
10. What role did animals have in Cherokee culture? What other stories about the Cherokee and animals can you find?
11. What do the ulunsuti crystals represent in the uktena stories? Why are they dangerous?
12. What is the relationship between snakes and birds? Where does this appear in other Indigenous North American stories, or worldwide?
13. Is the snake growing from Aganunitsi's hair related to the Haudenosaunee story of the Peacemaker, where the Tadodaho had snakes tangled in his hair during his evil days?
14. What role does water play in these stories, and in Seminole culture?
15. What accounts for the variety of snake legends among the Seminole and culturally related groups?
16. In several stories, people were transformed into snakes. How do these transformations differ from animal transformations in other Indigenous stories?
17. In this story, the Ouroboros image appears when Kolowissi must put his own tail in his mouth. Why does this image appear in multiple cultures worldwide? What does it signify here?
18. What do these stories tell us about how the Zuni conceive of human relationships, and responsibility to others?
19. One of the main tenets of Hopi culture is respect and understanding for others, including animals and the land. Do these stories help develop a sense of empathy? Who do you empathize with in these stores?

20. What role does forgiveness and punishment have in these stories?

21. One of the most famous Aztec (or Nahua) stories involves another major god, Huitzilopochtli, dismembering his sister Coyolxauhqui and flinging her body down Coatepec mountain, a scene that the Aztec recreated during festivals by practicing human sacrifice on the main temple of Tenochtitlan, their capital city. What parallels do you see in this story?

22. Across "Mesoamerican" Indigenous cultures, Quetzalcoatl has taken on different traits to different cultural communities. In some, he was a kindly god who encouraged trade and learning. In others, he calls for human sacrifices. What traits does he have in this story? Why do stories about him differ?

CHAPTER 9: NORSE

Background:

The Scandinavian Homelands stretch from Norway south through Denmark and west from, at least, Greenland through to western Russia. In this chapter, we include the alpine region of central Europe. As much as possible, the myths are presented in their oldest form, however, the introduction of Christianity has likely influenced some of the stories.

Dragonet of Mt. Pilatus

Background:

Mt. Pilatus is a vast mountain with several peaks ranging from 6,982 feet to 6,906 feet. Fossils of pterodactyls and archaeopteryx have been found throughout the region around Mt. Pilatus. A popular account of a dragon sighting on the mountain, written by Athanasius Kircher in the 1600s, is often cited in travel guides for the region.

Myth:

Stories are still told in the area of Mt. Pilatus, Switzerland of a dragonet, a dragon about the size of a human, but still potentially lethal. Its blood alone is said to kill a person with a single drop and its breath produced a poisonous cloud.

Many, many years ago, at least as long ago as the medieval times, near the town of Wilser, near Mt. Pilatus in Switzerland, there was a dragonet who was terrorizing the farms and folk. It killed and ate whatever it wanted and even though it was not a large creature like most dragons, none were able to slay it.

There was a man named Winckelriedt, most say he was an aggressive warrior known for misdeeds and banished from the town. The chronicle of Petermann Etterlein says Winckelriedt was the Regional Governor, but most say he had killed a person and been banished. In the town's time of need, they asked Winckelriedt to come back and slay the dragon. If he slayed the dragon, he would be allowed to stay.

So, Winckelriedt climbed the mountain to the dragon's lair armed with his mighty sword. The dragon did not wait for Winckelriedt to come close, but charged him ferociously. The battle ensued, and at one point, the dragonet came close enough for the warrior's sword to slice his neck and kill him. Winckelriedt raised his sword in triumph, only to have the dragonet's poisonous blood drip down upon the warrior and kill him.

There is, however, another story about dragonets and Mt. Pilatus. It is recorded that in 1421, a farmer, Stempflin had a close encounter with a dragonet on the mountain.Stempflin was so frightened that he passed out. When he awoke, he found a pile of dried blood and a dragon stone. In 1509, the stone was "confirmed" to have healing powers.

Illustration:

Archaeopteryx, the famous transitional fossil between non-avian dinosaurs and modern birds, was first found in limestone deposits in southern Germany. This is not far from where the story of the Dragonlet of Mt. Pilatus takes place. This feathered dinosaur was the inspiration for this illustration.

SOURCES:

Ashton, John. (1890, 2014.) Curious Creatures in Zoology. HardPress Publishing.

Bane, Theresa. (2015.) Encyclopedia of Beasts and Monsters in Myth, Legend, and Folklore. Jefferson, North Carolina, USA: McFarland & Company, Inc.

Hargreaves, Joyce. (2009.) A Little History of Dragons. New York: Walker & Company.

Mosby, Kristina and Jake Hicks. Mount Pilatus. Lucerne Switzerland. Retrieved from https://lucerneswitzerland.weebly.com/mount-pilatus.html.

Rose, Carol. (2000.) Giants, Monsters, and Dragons: An Encyclopedia of Folklore, Legend, and Myth. New York, USA: W. W. Norton & Company, Inc.

Shuker, Karl. (1994.) Dragons: A Natural History. New York: Barnes & Noble Books.

Fáfnir

Background:

Originally found in a set of poems called the Elder Edda as well as the Saga of the Volsungs, written in the thirteenth century B.C./B.C.E. by an unknown author, this story features the Volsung clan, and a family feud gone terribly awry. The Volsung Saga has three major parts, and the second portion covers Sigurd and the destruction of the dragon, Fafnir.

Myth:

The myth revolves around a journey wherein the gods Loki, Hœnir, and Odin came across an otter eating a salmon. The Otter was actually one of three sons of Hreithmar. Not knowing this, they killed him and brought his otter body to his father Hreithmar's house. Hreithmar recognized the otter as his son, and with his other sons Fáfnir and Regin took the gods hostage. As repayment for his son's life, Hreithmar demanded that Loki go out and fill the otter's skin with gold and to then cover the skin in gold.

Loki, ever the cunning trickster, went to the dwarf Andvari to collect the gold. Andvari gave Loki cursed gold, as well as the cursed ring Andvaranaut. This ring and the cursed gold were said to bring death to whoever owned them. Loki passed these things to Hreithmar and thus freed Hœnir and Odin. Seeing his father's wealth, a great greed overtook Fáfnir and he murdered his own father. As his greed grew, he fled further into the wilderness with his ever-growing horde of treasure. His body twisted into serpentine features, until finally he was changed into a terrible dragon. His breath was poison to the land, and he spread desolation far and wide around him so that no one would come near his treasure.

Fáfnir's brother Regin had not forgotten Fafnir's original act of betrayal, and sought vengeance on his brother for the death of their father. But beneath his rage and loss, Regin also considered that if Fáfnir were to die, he might claim all the treasure for himself. So Regin sent his son Sigurd to slay the dragon, giving him instructions to dig a pit and wait for Fáfnir to crawl over him on his way to the river where he went to drink every day. Sigurd dug the pit and waited with his sword Gram, but Odin came to him as an old man. Odin warned him to dig many more pits connected to the first, for if he did not, the blood from Fáfnir might cover and drown him. Sigurd followed these instructions, and it was not long before the earth shook with Fáfnir's footsteps. Sigurd waited and thrust Gram into Fáfnir's shoulder as he passed overhead. As he lay dying, Fáfnir asked Sigurd who he was and who had sent him on his mission. Discovering that Sigurd was his nephew, he warned him that Regin would be his doom, and that the treasure Sigurd sought for himself and his father would bring only death and misery. Sigurd disregarded Fáfnir's advice, and soon presented his father, Regin, with both the dragon's great treasure horde and his heart.

Yet Fáfnir was not entirely wrong to doubt the noble intentions of his brother Regin. Regin asked Sigurd to cook Fáfnir's heart for him, and all the while thought of killing his own son so that the dragon's treasure would belong to only him. Yet as Sigurd cooked the heart, he digested a bit of Fáfnir's blood which gave him the ability to understand the language of birds. Listening in wonder to their speech, he heard them describe his own father's plot to kill him. Sigurd took the sword Gram and took his father's head off. He ate Fáfnir's heart himself, saving a part of the heart which he later gave to his wife, Gudrun.

The saga continues on, changing its focus to Gudrun and her Burgundian family.

Illustration:

Fafnir is as iconic as he is nondescript. Like many European dragons, he is not described in detail. Here, he is drawn with bat wings, goat horns, and a snake-like body as he appears in many historical depictions.

SOURCES:

Aldington, Richard and Delano Ames, Transl. (1959.) New Larousse Encyclopedia of Mythology. Hong Kong: Prometheus Press.

Byock, Jesse L., trans. (1990.) The Saga of the Volsungs: The Norse Epic of Sigurd the Dragon Slayer. Berkeley and Los Angeles: University of California Press.

Creedle, William. (2010.) The Otter's Ransom: Moral Accompaniments to Legal Codes in the Icelandic Sagas. Lulu Press, Inc.

McLeish, Kenneth. (1996). Myth: Myths and Legends of the World Explored. New York, NY: Faces On File, Inc.

Schorn, Brittany. (2018.) Epic Tales: Norse Myths and Tales. London, UK: Flame Tree Publishing.

Somerville, Angus A. and R. Andrew McDonald, Eds. (2010.) The Viking Age: A Reader. Toronto: University of Toronto Press.

Gesta Danorum Dragon

Background:

Many histories of Denmark and Scandinavia have been passed down through the ages. The Gesta Danorum (Deeds of the Danes) is a compilation of some of these stories written down in the 12th century by Saxo Grammaticus. It contains nine books that focus on Norse mythology.

Then there are seven more books which appear to describe historically documentable events and kings.

Myth:

For at least twelve years, a large and fearsome monster, Grendl, had harassed the kingdom of King Hrothgar. Some equate Grendl with vampires, some with dragons. The monster tended to mostly hunt at night, usually very close to the mead hall of the king, Hrothgar. Grendl despised hearing the king's court enjoying a night of revelry. One particular night, in a fit or rage, Grendl descended upon the king's hall and grabbed the first man he encountered and tore him apart and ate him. Some say he continued this and slaughtered up to thirty men. The court moved away, trying to find a safer place. However, Grendl continued to find the court and his rampage lasted for years.

After many years went by, a young warrior intent on proving his greatness came to King Hrothgar's court. His name was Beowulf. He offered to rid the king and everyone of the monster. The usual festivities took place and Grendl came to devour his fill. However, this time, the warrior, Beowulf came to the court's rescue. Beowulf fought Grendl, tearing off his arm which ultimately ended in Grendl's death. The arm was hung in the mead hall as a trophy.

Grendl managed to crawl off to his home and die there, next to his mother. Yes, Grendl had a mother with whom he shared the territory. His mother was furious. Not only did her son die, but the murderers had hung his arm as a trophy. She went in search of revenge. She entered the hall, gained possession of her son's arm and grabbed and ate one of the king's men.

The myth continues, saying that the next morning, Beowulf travelled to the home of Grendl and his mother to fight the mother. Some say it was a vast swamp and Beowulf had to swim deep to enter the lair of the mother and son. They say that it was a magical place protected by the witchcraft of the mother. But Beowulf found a magical sword there and was able to overcome the mother and save the kingdom.

Many years later, when Beowulf was a king of his own territory, and an aging warrior, yet another battle with a dragon took place. A servant of Beowulf, unhappy with his lot, stole away to a barrow, a place sometimes known to hide monsters and treasures. There, in this place that had been undisturbed for centuries, the servant found great treasure. There were stories that the treasure had been abandoned by an ancient elite warrior. The servant thought it would be a fine idea to take a gold goblet. He didn't know that after the treasure had been abandoned, a mighty dragon had taken residence and considered the treasure to be his own. He was furious that some mere human had stolen even the smallest piece of his treasure.

Once more, Beowulf must go forth and battle, this time a fire breathing monster. Whether Beowulf told his followers to stay back or they stayed back from fear, Beowulf fought the dragon alone. However, one warrior and kinsman, Wiglaf, finally came to the aid of the king and together they were able to stab the beast in its underbelly. But Beowulf was mortally wounded and he was buried with the treasure that the dragon coveted for oh so many years.

Illustration:

The Gesta Danorum Dragon is typical of many European dragons; it is described as a large reptilian beast with wings. Here, patterns from the green-scaled lizard *Lacerta agilis* and bat wings are used to bring this dragon to life.

SOURCES:

Ashliman, D. L. (2010, October 26.) Beowulf: A summary in English prose. Retrieved from www.pitt.edu/~dash/beowulf.html#three.

Bane, Theresa. (2015.) Encyclopedia of Beasts and Monsters in Myth, Legend, and Folklore. Jefferson, North Carolina, USA: McFarland & Company, Inc.

Evans, Jonathan. (2008.) Dragons. London, UK: Apple Press.

Hall, Lesslie, trans. (1892.) Beowulf: An Anglo-Saxon Epic Poem, Translated From The Heyne-Socin. Boston, New York, Chicago: D. C. Heath & Co., Publishers.

Hargreaves, Joyce. (2009.) A Little History of Dragons. New York: Walker & Company.

Ingersoll, Ernest. (1928, 2014.) Dragons and Dragon Lore. New York: Cosimo Classics.

Niles, Doug. (2013.) Dragons: the Myths, Legends, & Lore. Avon, Massachusetts: Adams Media.

Rose, Carol. (2000.) Giants, Monsters, and Dragons: An Encyclopedia of Folklore, Legend, and Myth. New York, USA: W. W. Norton & Company, Inc.

Jormungandr

Background:

The Prose Edda is the classic literature describing Norse Mythology. It is likely that these stories were first recorded in the Elder Edda, and then written down in the 13th century Prose Edda, likely by Snorri Sturluson. Sturluson was an Icelandic politician and poet. The Prose Edda contains the Old Norse Poetics. Sturluson was a Christian and some say his religion influenced his telling of the Prose Edda. It is difficult to accurately gauge the age of the stories, possibly they were passed down orally for hundreds of years before Sturluson wrote them down.

Most scholars of mythology agree that Jormungandr represents chaos at the creation of the world, and will also bring its destruction through chaos. Some ancient texts state that Jormungandr and Nidhog/ Nidhogger are the same serpent, and that Nidhog's assault on the tree of life, Yggdrasil, is another chaotic behavior; others have them as entirely separate beings.

Myth:

The Norse Saga is full of many tales of Jormungandr and Thor. One element remains constant in all of the stories: Thor and Jormungandr were mortal enemies.

Jormungandr was one of three children of Loki and the giantess, Angrbova. The three children were Jormungandr, Hel (goddess of the underworld) and the ferocious Fenfir (a wolf that few could approach). The gods were appalled when they saw Jormungandr. He was huge and ugly, spitting venom, so they threw Jormungandr into the ocean where he grew even larger and more horrific. He grew into a very large ouroboros type creature. His tail was in his mouth and he encircled all of our world, known as Midgard.

Twice, Thor and Jormungandr faced off before Ragnarok. The first was when Thor decided that he really wanted to go fishing with the giant, Hymir. Thor didn't have any bait, so Hymir told Thor to find some. Thor found Hymir's herd and cut off the head of the largest ox to use for bait. Then Hymir and Thor went fishing in the ocean. They caught two whales, but that was not enough for Thor, so they went even further into the ocean. Thor put the ox head on his line and got a tremendously strong pull. It was Jormungandr. A great tug of war ensued, but when the serpent tired and Thor got him close to the surface, Hymir had an anxiety attack; so afraid of Jormungandr was he, that he cut the line and the serpent fell back into the ocean.

The second encounter was when Thor went to visit the giant king, Utgardhaloki. The giant king was not fond of the gods. He felt that they regularly disrespected him, so when Thor came to visit, the giant king set three challenges for Thor. The first challenge was to pick up the king's cat off of the floor. Thor, a very mighty warrior, could only lift one of the cat's paws up off the floor. The king laughed loudly.

The second challenge was to drink a horn of mead in three gulps. Thor tried as hard as he could, but could not empty the horn. The third challenge was to wrestle the king's old nurse. Again, Thor was defeated. The king escorted Thor out of the kingdom, and only when Thor was in an area where he could not harm the kingdom, did the giant king explain the tricks. The cat was Jormungandr, the horn of mead was connected to the ocean (some say that Thor's drinking caused the first ebb tide) and the old nurse was old age. Before Thor could release his terrible temper, the giant king disappeared!

The final encounter, the Day of the Last Battle, Ragnarok, will see a fierce battle between the mortal enemies. Jormungandr will rise from the water, churning the ocean to its floor and spilling water over the land. He will join his father, Loki, and his siblings in attacking Asgard. As he spews poison, Fenrir the giant wolf will spew fire, and both the sun and moon will be devoured by wolves. In the end, Thor will be able to crush the serpent's head and kill him, but with Jormungandr's final poisonous breath, he will kill Thor. A new world will be born from this destruction. In some tellings, Ragnarok has already taken place; in others, it is a future yet to come. This may be due to a belief in fate - that even if the battle has not happened yet, its full details can already be known.

Illustration:

Jormungandr, the great serpent of the sea, is inspired here by the extinct *Dunkleosteus* and extant fin whale (*Balaenoptera physalus*). Dunkleosteus could grow up to 20 feet long and was covered in bony armor that also functioned as its teeth. A special joint in its head allowed it to have an extra large gape.

SOURCES:

Densmore, S. M. (1976). Mythic Allusion in DH Lawrence's women in Love. Doctoral dissertation. Retrieved from https://macsphere.mcmaster.ca/bitstream/11375/9561/1/fulltext.pdf.

Bane, Theresa. (2015.) Encyclopedia of Beasts and Monsters in Myth, Legend, and Folklore. Jefferson, North Carolina, USA: McFarland & Company, Inc.

Evans, Jonathan. (2008.) Dragons. London, UK: Apple Press.

Gordeev, Nikolai P. (2017.) Snakes in the Ritual Systems of Various Peoples. Anthropology & archeology of Eurasia 56(1-2), 93-121.

Jormungand. (2016.) Norse Mythology. Retrieved from http://norse-mythology.org/gods-and-creatures/giants/jormungand/.

McCall, Gerrie and Regan, Lisa. (2011.) Monsters and Myths: Dragons and Serpents. New York, NY: Gareth Stevens Publishing.

McLeish, Kenneth. (1996). Myth: Myths and Legends of the World Explored. New York, NY: Faces On File, Inc.

Niles, Doug. (2013.) Dragons: the Myths, Legends, & Lore. Avon, Massachusetts: Adams Media.

Rose, Carol. (2000.) Giants, Monsters, and Dragons: An Encyclopedia of Folklore, Legend, and Myth. New York, USA: W. W. Norton & Company, Inc.

Schorn, Brittany. (2018.) Epic Tales: Norse Myths and Tales. London, UK: Flame Tree Publishing.

Shuker, Karl. (1994.) Dragons: A Natural History. New York: Barnes & Noble Books.

Nidhogg

Background:

The Prose Edda is the classic literature describing Norse Mythology. As mentioned above, it was probably written down in the 13th century, by Snorri Sturluson. Nidhogg is a major force in the competition for order. Some sources say that Nighogg represents the afterlife, but others claim that this is the result of Christian influence on the story. Others say he is a representation of winter. In most stories, he is explicitly described as a winged serpent.

Myth:

The Norse saga that ends in Ragnarok is a story that has long been told. There is a great ash tree, Yggdrasil, that runs through and connects the nine levels of the world. At the top sits a great eagle, who proudly overlooks the world.

At the bottom of the tree is Nidhogg, a great serpent who chews on the roots of Yggdrasil. Some myths say that he also eats dead flesh and drinks blood, sometimes cooking corpses in Hvergelmir, a bubbling cauldron. Occasionally his name has been translated to "corpse-tearer," though Nidhogg has many translations, the most common one being "dread biter". The root of Yggdrasil is known as Nifhein, sometimes translated as the realm of the dead. When he has time, he send taunts to the eagle at the top of the tree, which Christianity refers to as Heaven. The messages are delivered via a giant squirrel named Ratatosk. Ratatosk also brings provocative messages back from the eagle. Some say that the roots of the tree are what keeps Nidhogg stuck at the bottom.

When Ragnarok happens (or happened, depending on the version), the saga says that Nidhogg will rise up on his wings and will take the dead up with him, on his wings, to fight in the final battle. Nidhogg will survive the final battle and become a presence in the next world.

Illustration:

Nidhogg lives underground and gnaws on the roots of the world tree. While more interested in eating worms than roots, the fire salamander (*Salamandra salamandra*) spends most of its time in the leaf litter beneath large trees. They are called fire salamanders because of their yellow pigments and their tendency to scurry out from burning wood that had once been a cool, wet home.

SOURCES:

Aldington, Richard and Delano Ames, Transl. (1959.) New Larousse Encyclopedia of Mythology. Hong Kong: Prometheus Press.

Bane, Theresa. (2015.) Encyclopedia of Beasts and Monsters in Myth, Legend, and Folklore. Jefferson, North Carolina, USA: McFarland & Company, Inc.

Evans, Jonathan. (2008.) Dragons. London, UK: Apple Press.

Jobes, Gertrude. (1962.) Dictionary of Mythology Folklore and Symbols. New York, NY: The Scarecrow Press, Inc.

Jormungand. (2016.) Norse Mythology. Retrieved from http://norse-mythology.org/gods-and-creatures/giants/jormungand/.

McCall, Gerrie and Regan, Lisa. (2011.) Monsters and Myths: Dragons and Serpents. New York, NY: Gareth Stevens Publishing.

McCormick, Kylie. (2013, September 9.) Nidhogg. Dragons of Fame. Retrieved from www.blackdrago.com/fame/nidhogg.htm.

McLeish, Kenneth. (1996). Myth: Myths and Legends of the World Explored. New York, NY: Faces On File, Inc.

Nidhogg. Draconika. Retrieved from http://www.draconika.com/legends/nidhogg.php.

Nidhogg. (2012, January 26.) tobiasmastgrave.wordpress. Retrieved from https://tobiasmastgrave.wordpress.com/tag/nidhogg/.

Niles, Doug. (2013.) Dragons: the Myths, Legends, & Lore. Avon, Massachusetts: Adams Media.

Schorn, Brittany. (2018.) Epic Tales: Norse Myths and Tales. London, UK: Flame Tree Publishing.

Rose, Carol. (2000.) Giants, Monsters, and Dragons: An Encyclopedia of Folklore, Legend, and Myth. New York, USA: W. W. Norton & Company, Inc.

Lagarfljotsòrmurinn

Background:

Stories are not only still told, but even caught on video, of the Lagarfljotsòrmurinn - murinn signifying "worm." This Icelandic lindworm is said to live in Lake Lagarfljot, a glacial fed, freshwater lake which exists below sea level. Sightings are as recent as 1963, 1983 and 1998. Adding to the mystery are accounts from experts that they have seen bones in the area that did not clearly correspond to local fauna. Some say that it was video taped in 2012, thought at least one scientist has asserted that the object is an icy fishing net. An Icelandic panel did declare this video footage the real Lagarfljotsòrmurinn, though their motives have been questioned as a possible tourism stunt. You can make your own decision about that.

Myth:

The earliest tales go back to 1345 and tell of a mother who gave her daughter a gold ring. The daughter asked how to increase the amount of gold. The mother told the daughter to place the ring under a lindworm inside a chest. The daughter did so, and after a few days checked on the growth of the gold. The lindworm had grown, but the gold had not. In fact, the lindworm was rapidly gaining the size of a full dragon and the chest had fallen apart. The daughter threw the lindworm and the chest into the lake, Lake Lagarfljot. Some say that these dragons keep growing as long as there is gold underneath them and when the mini-dragon was thrown into the lake, the gold went with it and settled under it.

After a while, local folk noticed that there was a very large serpent type creature in the lake and it was starting to devour local livestock and people. Some said it was capable of spitting venom. The people in the countryside were very afraid, but had no solution.

Most of the versions of the myth tell of two very brave Finnish men who were asked to come and kill the dragon. The two men went down into the lake for a very long time. When they surfaced, they said that they were unable to kill the serpent, but were able to tie its head and its tail to the bottom of the lake. However, they said that there was an even bigger dragon with an even bigger hoard of gold under the younger dragon.

The Lagarfljotsòrmurinn is a harbinger of doom, and some say that if it ever gets free of the bindings, it will bring about the end of the world.

Illustration:

The european eel (*Anguilla anguilla*) can grow up to 4 feet long and is catadromous, meaning it lives in freshwater and travels to saltwater to breed. This serpentine freshwater fish may explain sightings of the lake monster Lagarfljotsòrmurinn.

SOURCES:

Baring-Gould, Sabine. (1863.) Iceland: Its Scenes and Sagas. London: Smith, Elder.

Jauregui, Andres. (2014, September 24.) Lagarfjótsormur, Iceland's Legendary Lake Monster, Caught On Tape, Panel Says. Huffngton Post. Retrieved from https://www.huffngtonpost.com/2014/09/24/lagarfjotsormur-sea-monster-video_n_5875970.html.

Simpson, Jacqueline. (1972). The Water-Snake of Lagarfjot. Icelandic Folktales and Legends. Berkeley and Los Angeles: University of California Press.

Lindworm

Background:

As far back as 4000 years ago, there are engravings of maned serpents on Nordic bronze razors. Some have horns. Some guard burial mounds. Others guard treasures, and still others guard maidens. It is a very, very large serpent with a horse's head and mane, bat-like wings, and some say it also has gold or greenish scales.

The stories come from all over Scandinavia, and Germany. Marco Polo even describd seeing lindworms on his travels, fierce enough to attack horses. Many stories hail from Sweden, where the following story is still told. Some spell the dragons as lindworm; others use lindorm. Here, we use lindworm.

Myth:

Long ago, there was a king and a queen who desperately wanted children, but had never had any. The queen decided to go to see a soothsayer and ask for advice. Some say the soothsayer was a beggar woman at the castle. Whoever she was, she told the queen that if she ate two onions then the queen would give birth to twins. One small problem, the queen was so excited that she ran off before hearing the full advice: peel the onions before eating. The queen ate the first onion without peeling. It tasted so awful that she peeled the second one.

So, nine months later, the queen gives birth to two sons, except the first one does not look human. It looks like that ancient monster, the lindworm. Some say the king threw it out of the window, some say it was the queen, some say it was the midwife. Whoever tossed the lindworm out of the window, the lindworm grew up in the forest, all alone. The second son grew up in the lap of luxury, well and truly loved and cared for.

When the second son reached adulthood, it was time to find a bride, but the lindworm kept getting in the way, stopping prospective brides from reaching the prince. When the prince went in search of the problem, he came upon the lindworm who quickly explained that he was the older brother, and by rights, should marry first and it must be a love match. But who would marry a lindworm?

Some say, the royal family offered the lindworm a slave woman every day for many days, but the lindworm refused. (Some say he tore those women into pieces). Finally, a village maiden who had the good sense to seek out the original soothsayer, was sent to the lindworm. When presented to the lindworm, the maiden was wearing a quantity of clothing. When the lindworm told her to undress, she struck a bargain. For every dress that the maiden removed, the lindworm had to remove a layer of skin, until she was naked.

The final layer on the lindworm disappeared into a mist. Some say the mist was from the maiden scrubbing the last layer off of the lindworm. However it happened, when the final layer was removed, there stood a strong, handsome man, the Lindworm prince.

He became the Lindworm King and ruled for a long time. And many other stories follow the exploits.

Illustration:

The Lindworm is described as a reptilian monster with a horse-like head and horns. Here, ibex horns, horse ears and hair, and patterning from the venomous snake *Vipera berus* were used to construct the Lindworm.

SOURCES:

Asbjørnsen, Peter Christen. (1977.) East of the Sun and West of the Moon: Old Tales from the North. Garden City, New York: Doubleday.

Bane, Theresa. (2015.) Encyclopedia of Beasts and Monsters in Myth, Legend, and Folklore. Jefferson, North Carolina, USA: McFarland & Company, Inc.

Dekirk, Ash. (2006.) Dragonlore: From the Archives of the Grey School of Wizardry. New Jersey: Career Press, Inc.

Lang, Andrew. (1897.) The Pink Fairy Book. Dover: Dover Publications.

Meurger, Michel and Claude Gagnon. (1998.) Lake Monster Traditions: A Cross-Cultural Analysis. London: Fortean Times.

Rose, Carol. (2000.) Giants, Monsters, and Dragons: An Encyclopedia of Folklore, Legend, and Myth. New York, USA: W. W. Norton & Company, Inc.

Shuker, Karl. (1994.) Dragons: A Natural History. New York: Barnes & Noble Books.

Drachenfels Dragon

Background:

Just south of Bonn in Germany is Drachenfels or "dragon rock." This is a hill formed by molten lava which did not break through the earth's crust, but stopped and cooled and is today Drachensfel. Local legends still focus on the draconic myths of the site, and this perhaps contributed to Drachensfel being the first protected natural monument in Germany.

Myths:

The stories go back many, many years and have been told again and again.

There are two distinct legends that surround this well known rock.

The first legend involves the saga about Fafnir (see Nordic Myths, Fafnir) and that Drachensfel is where Sigurd killed the dragon, Fafnir and gained invulnerability by bathing in Fafnir's blood. The Nordic version says that Sigurd ate a piece of Fafnir's heart and was then able to hear the words of the birds who told him of his father's greed and intent to kill Sigurd. Is this a form of invulnerability?

Also, in one of the Nordic versions, Sigurd has a magic sword, re-forged by Sigurd and Regin. In this Germanic version, Sigurd is a blacksmith, so strong he was feared and sent into the forest to kill the dragon, Fafnir. Sigurd was not expected to be successful and return. But Sigurd's sword is so strong that he is able to kill the dragon, Fafnir. Sigurd bathes in Fafnir's blood and becomes invulnerable.

Many, many years later Drachenfels is the scene of another dragon legend. This time, a non-Christian tribe (some would say pagan, others say heathen), lived at the bottom of the hill still known as Drachenfels, Dragon Rock. Two of their mightiest warriors were brothers, Polterich (alias Horsrik) and his younger brother Rauferich (alias Rinbod). These two frequently led raids onto the other side of the Rhine

River where a Christian folk now lived. In one of these raids, they captured several people including a beautiful young maiden. Both brothers wanted the maiden for his own, though on account of their religious differences she wanted neither of them. They had a vicious fight, so violent that the eldest of the clan declared that neither brother would have her. Instead, she would be sacrificed to the dragon who lived on the mountain, Drachenfels.

The myth says that this dragon was a most monstrous beast who spit fire from its mouth and eyes. Now, there are two versions to what happened next.

One story says that the dragon was not actually a fire breathing dragon but an old Dragon lady who Rauferich, the younger brother, had met during one of his adventures. When the elder declared that the beautiful young woman would be sacrificed to the dragon, Rauferich went to the old Dragon lady and they decided to put on a show. The sacrifice was sent to the dragon wearing a golden cross. When the dragon (Dragon lady, those below apparently couldn't tell the difference) came close to the woman, it saw the cross, let out a terrible roar heard all around the valley, burned a hole in the side of the mountain and disappeared never to be seen again (the plan was for the Dragon lady to take a long warm nap, say for centuries).

A second, more recent version, says that it was a pagan priest who declared the Christian virgin be sacrificed and he, himself tied her to the sacrificial rocks to be devoured by the dragon. The Christian virgin awaited death, praying constantly and when the dragon approached, she managed to bring forth the golden cross from around her neck and hold it up for the dragon to see. Then, the dragon rose up, frightened, and threw itself on the jagged rocks in the water below, never to be seen again.

All of the pagan folk were so astonished that they released the maiden and the other prisoners that they had. Later, Rauferich and the old priest converted to Christianity, soon followed by all of the other pagans.

Illustration:

Drachenfels Dragon is typical of many European dragons, bat-winged and breathing fire. This one sports the horns of the alpine ibex (*Capra ibex*) and recoils away from the sight of the cross.

SOURCES:

Exploring Legendary German History on Dragon Rock. (2007, August 2.) Retrieved from http://www.dw.com/en/exploring-legendary-german-history-on-dragon-rock/a-2715873.

Mannering, May. (1866). Drachenfels; Or, The Dragon's Rock. Student & Schoolmate: An Illustrated Monthly for all our Boys & Girls 17(6), 218-221.

McLeish, Kenneth. (1996). Myth: Myths and Legends of the World Explored. New York, NY: Faces On File, Inc.

Niles, Doug. (2013.) Dragons: the Myths, Legends, & Lore. Avon, Massachusetts: Adams Media.

Portal, Claire. (2013.) When Unremarkable Landscapes Receive Heritage Status: Considerations Based on Case Studies in the Pays de la Loire (France). L'Espace géographique 42, 213-226.

Ruland, Wilhelm. The Drachenfels. Retrieved from http://www.kellscraft.com/LegendsRhine/legendsrhine081.html.

The legend of Drachenfels (Dragon Rock). (2015, November 19.) Middle Europe. Retrieved from www.middle-europe.cz/the-legend-of-drachenfels-dragon-rock.

The Legend of the Drachenfels; Or, How the Fell Dragon Fell. (1874). Magenta 4(1), 9-9.

Cultural Questions:

1. What lesson can be garnered from the death of Winckelreidt in the Mt. Pilatus dragon myth? Does this lesson relate to any other myths in the chapter?
2. Fafnir is a complex story where the dragon represents greed and other human emotions. Discuss whether the story is a cautionary tale and how it relates to the cultural system that it comes from.
3. What is the role of poison in relation to these dragon myths? Where else have you seen dragons associated with poison or venom?
4. What is the significance of Jormugandir with its tail in its mouth encircling the earth?
5. How does the death of the hero in Jormugandir impact the myth?
6. Within their own communities, equitable redistribution of wealth was often upheld by Norse cultural standards and greed was highly discouraged. How are these principles expressed through their myths?
7. What role does gold or treasure play across the myths?
8. What is your interpretation of the insults exchanged in Nidhogg?
9. Was the Lindworm justified in claiming a bride for himself before his brother? Relate this myth to other myths which focus on familial conflict.
10. How is revenge expressed across these myths?
11. What are the properties of blood in these myths?
12. Discuss the contradiction in the Mt. Pilatus dragon being able to heal or kill with poison.
13. Many modern authors, including J.R.R. Tolkien, have taken inspiration from Norse mythology. Can you see this influence in modern popular dragon lore?

CHAPTER 10: SLAVIC

Background:

The Slavic geographic area covers territory in Central, Eastern and Southeastern Europe as well as Northern and Central Asia. It is considered by many to be the Homeland of the Indo-European language family. Diverse in cultural developments, some cultural groups are still strongly connected while others regularly are in confrontation with each other.

It has been a crossroads of people and power over the millenia.

One strong characteristic of the Slavs in general is reverence for Mother Earth. This is the area of the earliest Venus figurines. Connections to Mother Earth reflect the agrarian societies throughout the area.

While Baba Yaga is not herself a dragon, her connection to dragons such as Chudo Yudo and Koshchei is symbolic of many aspects within the various cultural systems. These myths are open to your own interpretation of the importance of women in this area.

Alkha

Background:

In ancient Siberia, an area split today between Russia and Mongolia, known as Buryatia, there are many myths still told of Alkha (or Alicha or Arakho, depends on pronunciation). This area is the Homeland of one of the largest Indigenous communities in Siberia/northern Mongolia. Political subjugation has fluctuated between the Mongols in the 13th century and Soviet/Russian authorities.. Today it is within the Buryat Republic, under Russian political authority. Slavic folk traditions co-exist with the Russian Orthodox church and Buddhism.

Myth:

Alkha is a gigantic black winged serpent who can blot out the entire sky. Alkha resides in the heavens where he occasionally emerges to try to eat the sun or the moon. Both the sun and the moon are, of course, too hot to swallow so Alkha must spit them back out. However, this does not keep him from trying again and again. The local folk thought to scare Alkha into spitting out the sun or moon by throwing rocks and making noise whenever they saw that a bite had been taken out of the sun or the moon.

The gods came up with another solution for Alkha's insatiable need to eat the sun and the moon. They cut his gargantuan body in half. The lower half fell to the earth. Therefore, whenever Alkha took a bite, it simply fell out of his upper body and the sun or the moon was restored. Sometimes, this took until the eclipse has passed, but it always happened eventually.

Illustration:

In this illustration, Alkha gnaws on a moon. Its design takes elements from the eurasian black vulture (*Aegypius monachus*) and other reptiles like snakes and lizards.

SOURCES:

Alkha. (2013, December 25.) God Checker. Retrieved from http://www.godchecker.com/pantheon/slavic-mythology.php?deity=ALKHA.

Bane, Theresa. (2015.) Encyclopedia of Beasts and Monsters in Myth, Legend, and Folklore. Jefferson, North Carolina, USA: McFarland & Company, Inc.

Jobes, Gertrude. (1962.) Dictionary of Mythology Folklore and Symbols. New York, NY: The Scarecrow Press, Inc.

Rose, Carol. (2000.) Giants, Monsters, and Dragons: An Encyclopedia of Folklore, Legend, and Myth. New York, USA: W. W. Norton & Company, Inc.

Bolla

Background:

Many myths are still told of a very large serpentine monster, Bolla or Bullar in Albania. The Republic of Albania is in the southwestern corner of the geographic Slavic area . The area has endured political colonization by the Greeks and the Romans and political control by the Ottomans, Italians, and Nazi Germany. In 1991, it became the Republic of Albania. The majority of the population are Muslims although there are a number of other religions present, including Catholicism and Eastern Orthodox.

Myth:

Bolla is a dragon with a long snake like body with four legs and small wings. Her favorite food is humans. She hibernates most of the year, awakening on St. George's Day, traditionally April 23 which is considered by some to be the beginning of summer, but in ancient Albania it was the New Year. When she awakens on that day, she eats the first human to walk nearby, then she goes back to sleep until the next year.

When Bolla reaches the age of twelve, she develops into a Kulshedra, an even more evil monster. Now she is a large female dragon with nine tongues, horns, spines, and much larger wings. Some myths claim she has seven to twelve heads and breathes fire from every mouth. Some claim she can also be in the form of a woman,an eel, a turtle, frog or salamander. Her main weapons are her urine and her milk which she uses to drown folk and sometimes these tools poison people.

Albanians blame her for droughts as she is known to control the water in wells and springs. Some stories say she has red woolly hair. Some say she lives in a cave near a swamp or other body of water. If you see water that is red in color, it is said that it has some of the Kulshedra's blood in it. Appropriate to a water dragon, the Kulshedra is said to come with storm clouds. Albanians say she is a storm demon.

Most stories say that the Kulshedra is evil and wicked requiring human sacrifice to release her grip on sources of water. In particular, damsels were sacrificed to assuage her anger. However some stories say that the Kulshedra has a sense of humor. Kulshedra was known to be a protector of the Earthly Beauty. One story says that there was a very handsome young man who was in search of the Earthly Beauty. Kulshedra was said to be a protector of the Earthly Beauty. But when the Kulshedra saw this very handsome young man, she decided he was too good looking to eat and tried to help him by sending him to her next eldest sister for help in reaching his goal. That Kulshedra was not sure where the Earthly Beauty was to be found, but also found the young man too pretty to eat, so she sent him to the eldest sister, Kulshedra who was able to help the young man find and claim the Earthly Beauty.

Other stories say that Albanians are concerned that Kulshedra will steal the sun or moon so they shoot arrows into the air to scare her away. Loud noises made with pots, pans, metal objects, even church bells will also scare off the beast.

Illustration:

The Bolla is described as winged, coarse-furred, multi-headed with horns and sometimes exposed breasts. For this, elements from the female goat, bat, and the european grass snake (*Natrix natrix*) were combined into a terrible monster.

SOURCES:

Bane, Theresa. (2015.) Encyclopedia of Beasts and Monsters in Myth, Legend, and Folklore. Jefferson, North Carolina, USA: McFarland & Company, Inc.

Bolla. The Albania Wordpress. Retrieved from https://theealbania.wordpress.com/2014/08/08/albanian-mythology/.

Elsie, Robert. (2001.) A Dictionary of Albanian Religion, Mythology and Folk Culture. Washington Square, New York: New York University Press.

Lurker, Manfred. (2015.) A Dictionary of Gods and Goddesses, Devils and Demons. London and New York: Routledge.

Prifti, Peter R. and Biberaj, Elez. (2019, February 22.) Encyclopaedia Britannica. Retrieved from https://www.britannica.com/place/Albania.

Rose, Carol. (2000.) Giants, Monsters, and Dragons: An Encyclopedia of Folklore, Legend, and Myth. New York, USA: W. W. Norton & Company, Inc.

Chudo Yudo, Zmey Gorynych/Zmaj Gorynych, Sarkany

Background:

The following group of myths need a general introduction as they follow stories of beasts directly related to Baba Yaga. The myths cover territory from northern Russia down into Bulgaria. The stories and the depictions of the beasts with many heads show an evolution and history of the cultures, and regions that each exists in. This is an area that has experienced Greek, Roman, and Mongolian influence. Today and for quite some time, Russia has dominated the area. Russia, Poland, Bulgaria, the Ukraine, Serbia, Bosnia, Croatia and Slovenia have tales about dragons.

Dobrynya is based on a real warlord who led the armies of Svyatoslav the Great. He is generally believed to be Vladimir the great's maternal uncle and Vlad's tutor, around the 10th century A.D./C.E.

Throughout the Slavic region,especially as you travel south, dragons take on not only great power and strength, but are also seen as benevolent. Maybe it's not so unusual that Dobynya reached an accord with Zmey.

Chudo Yudo

Myth:

In many stories from Russia and nearby areas, Chudo Yudo is the name of an extremely large dragon, usually with multiple heads, 3 or 6 or 9 or 12, who can breathe fire. Some say Chudo Yudo is a descendant of Baba Yaga, the Slavic pagan goddess or the unmarried old woman who flies through the air in a mortar with a pestle for a paddle and a broom to sweep away her trail. Some say she was benevolent, some say the classic evil witch. Some say she had three daughters, two granddaughters. Some say she had many children.

Some say Chudo Yudo had a brother, Koshchei the immortal. Many consider Koshchei to be the embodiment of winter. As Winter, Koshchei imprisons spring, sometimes represented as a young woman, sometimes as a mother.

What follows is one version of a Chudo Yudo tale.

A long time ago, there was a family with three sons, the youngest, was Ivan Buikovich (Bull's son). One day Ivan realized that the land that he lived in was oppressed by a gargantuan Chudo Yudo who kept the land constantly in darkness. Ivan decided that when he got the chance, he would rid the land of the evil Chudo Yudo.

It came to be that one night, Ivan had the night watch over his brothers. While his brothers slept, a 6 headed Chudo Yudo rode up on his horse and Ivan fought and killed it. The next night, a 9 headed Chudo Yudo rode up. It was much harder to kill, but Ivan managed to finally kill it. On the third night, a 12 headed Chudo Yudo rode up on a horse with 12 wings and its coat was made of silver and its mane and tail were made of gold. Ivan managed to cut off three of the heads, but the Chudo Yudo was able to reattach the heads with its own "fiery fingers" (Ralston).

At this point, Ivan was up to his knees in the earth so he threw one of his gloves at the hut where his brothers were sleeping. The glove smashed through the window, but the brothers did not wake up. Then Ivan cut off six of the heads, but they grew back. Now, Ivan was half buried by the earth, so he threw his other glove at the hut. It pierced the roof. Still no one woke up. Ivan got buried up to his armpits fighting the Chudo Yudo so he threw his cap at the hut where his brothers were sleeping. The hut was blown apart. Ivan's brothers finally woke up and helped Ivan to destroy the Chudo Yudo.

However, that is not the end of the tale. The Chudo Yudo's widows decided to seek revenge on Ivan. Ivan shapeshifted into the form of a sparrow and overheard their plans. Each widow planned to take the form of something different, first a meadow with silk cushions, then an apple tree with golden fruit and then a spring, but Ivan knew the plan. As Ivan and his brothers tried to make their way home, they encountered each widow in their transformation, one at a time. But Ivan wouldn't let his brothers stop and rest on the cushions or stop to take an apple or to take a drink from the spring. He smashed each widow with a mace that his father had made him and killed them.

But the tale is still not finished. The mother of the widows was very angry and sought revenge. Some say she was the old witch, maybe Baba Yaga herself. She posed as a poor woman begging for a little help. Ivan gave her a coin, but the old witch grabbed his hand and dragged him off to her husband, the Aged One, in an underground dwelling. He could have killed Ivan, but instead he sent Ivan on an errand and a new set of adventures!

Illustration:

Somehow, the notably large and multi-headed Chudo Yudo is capable or horseback-riding. Here he is drawn with the patterning of *Vipera kaznakovi1*, riding what must be the strongest horse known to man.

SOURCES:

Bane, Theresa. (2015.) Encyclopedia of Beasts and Monsters in Myth, Legend, and Folklore. Jefferson, North Carolina, USA: McFarland & Company, Inc.

Chudo Yudo. Mythology Dictionary. Retrieved from www.mythologydictionary.com/baba-yaga-mythology.html.

Hubbs, Joanna. (1993.) Mother Russia: The Feminine Myth in Russian Culture. Bloomington and Indianapolis: Indiana University Press.

Johns, Andreas. (2004.) Baba Yaga: the Ambiguous Mother and Witch of the Russian Folktale. Peter Lang.

Jobes, Gertrude. (1962.) Dictionary of Mythology Folklore and Symbols. New York, NY: The Scarecrow Press, Inc.

McCormick, Kylie. (2003, September 9.) Chudo Yudo. Dragons of Fame. Retrieved from ww.blackdrago.com/fame/chudoyudo.htm.

Ralston, William Ralston Shedden. (1880.) Russian Folk-tales. R. Worthington.

Rose, Carol. (2000.) Giants, Monsters, and Dragons: An Encyclopedia of Folklore, Legend, and Myth. New York, USA: W. W. Norton & Company, Inc.

Warner, Elizabeth. (2002.) Russian Myths. Austin, TX: University of Texas Press.

Zmey Gorynych

Myth:

Usually the Zmey Gorynych is described as a very large winged dragon with three heads, green with two back legs and two smaller front legs. And, it can spit fire. Some say it is the descendant of actual dinosaurs. Whatever the truth is, Zmey Gorynych is known throughout the area as mostly mean and nasty, but sometimes benevolent, and always male and lusting after young women.

Some myths say he is an associate of Baba Yaga (see Chudo Yudo).

Zmey Gorynych means “snake”. Gorynych is thought to mean either mountain or burning, appropriate as the Zmey was known to live in caves on mountains and breathe fire. Some of the stories resemble those of Chudo Yudo, but one is very different.

Oral stories (bylinas) say that one time, Zmey Gorynych captured the niece of Prince Vlasimer of Kiev. A hero, (bogatyr), Dobrynya Nikitich, one of three adventuring brothers/heroes, took on the task of killing Zmey Gorynych and rescuing the princess, Zabara Putyatishna. Dobrynya had encountered Zmey once before, but unusual in Slavic folklore, they had reached an agreement not to kill each other. Dobrynya went in search of the princess, into the Saracen mountains, where an epic battle took place for three days and three nights. Finally, Dobrynya cuts off all three heads and ends the life of Zmey Gorynych, thus rescuing the princess and freeing ancient Rus from the oppression of Zmey Gorynych.

Some versions say that since Dobrynya was a peasant. He could not marry the princess, so she married Aloysha Popovich.

Some stories say Aloysha was Dobrynya's brother by oath (knightly oath) and that there is a whole other story about Dobrynya already being married to another woman, Natashya, but that Dobrynya had to go on a 12 year quest. He told Natashya that if he did not return after 12 years, she was to marry someone else. But Aloysha Popovich kept telling the court and Natashya that Dobrynya was already dead and that she should marry Aloysha. Finally, the Prince believes Aloysha and orders Natashya to marry Aloysha, but on the wedding day, an amazing minstrel arrives who so entrances Natashya that she asks him to come and sit by her whereupon she realizes that it is her own dear Dobrynya who then gives Aloysha a good thumping.

There is one more myth about Aloysha Popovich and the Tugarin Zmeyevich. The most common version has it that Tugarin Zmeyevich was at a feast at Prince Vladimir's. Tugarin insulted the Prince and his wife, didn't pray to God and was a glutton. Aloysha told a story intended to insult Tugarin, so Tugarin threw a knife at Aloysha. This called for a duel. When Aloysha and Tugarin meet, Tugarin is flying on fragile, paper thin wings, but using magic to pummel Aloysha with fire and smoke. Aloysha prays for rain which soaks and destroys Tugarin's wings. Then Aloysha cuts the dragon into pieces and cuts off the dragon's head which he presents to the prince on a pike.

Zmaj Gorynych

Myth:

Zmaj Gorynych is closely related to Zmey as can be seen in the following tale. Zmaj seems to be much like Asian dragons, often benevolent. Usually they are said to have a ram's head and a large serpent's body. They are said to protect humans from the Ala or Azjada, a bad weather demon. Many of the myths are from Serbia.

Many legends say that Zmaj loved to take a human form and pursue women, much like Zmey. If bad weather continues for any length of time, folk take action to chase the Zmaj out of their homes.

One story even claims that a Zmaj was the lover of a tsarina for a year. The tsar, said to be Tsar Lazar, discovered that the Zmaj could only be scared off by the Zmaj-Despot Vook so the tsar asked the Zmaj-Despot Vook to come. Vook came and slayed the Zmaj lover of the tsarina.

Some say the Zmaj-Despot Vook is based on an actual historical figure, Despot Vuk Brankovic, who enjoyed being the hero of this story and even claimed to be a descendant of a dragon. Many Serbian rulers employed myths of Zmaj to increase their reputation as extremely strong rulers.

Illustration:

Lacerta viridis, the European green lizard gives this Zmey its legs and color, but the rest of the design is inspired by classical renditions of European dragons. Crowns and spade-tipped tails are traditional.

Sarkany

Myth:

Hailing from Hungarian mythology is the legend of the Sarkany. Born the son of Boszorkany, a terrible witch, (could this be Baba Yaga?) Sarkany is described as a man-like dragon demon with multiple heads—often three, seven, or twenty-one. He has dominion over thunder and storms, and is usually seen riding a horse. In battle, he wields a morning star and a sword and possesses the ability to turn his foes to stone.

Sarkany definitely has a strong resemblance to Chudo Yudo and Zmey.

Illustration:

Sarkany (see Chudo)

SOURCES:

Allen, Peter J. and Chas. Saunders., Eds. (2013, January 18.) Zmey Gorynich. God Checker. Retrieved from http://www.godchecker.com/pantheon/slavic-mythology.php/deity=ZMEY-GORYNICH.

Bane, Theresa. (2015.) Encyclopedia of Beasts and Monsters in Myth, Legend, and Folklore. Jefferson, North Carolina, USA: McFarland & Company, Inc.

Chudo Yudo. Mythology Dictionary. Retrieved from www.mythologydictionary.com/baba-yaga-mythology.html.

Dixon-Kennedy. (1998.) Encyclopedia of Russian and Slavic Myth and Legend. Santa Barbara, California: ABC-CLIO.

Dobrynya Nikitich and Zmey Gorynych. (2016.) Ural Stone Carving. Retrieved from stonecarving.ru/dobrynya-nikitich-and-zmey-gorynych.html.

Georgieva, Ivanička (1985). Bulgarian Mythology. Translated by Vessela Zhelyazkova. Svyat Publishers.

Hubbs, Joanna. (1993.) Mother Russia: The Feminine Myth in Russian Culture. Bloomington and Indianapolis: Indiana University Press.

Johns, Andreas. (2004.) Baba Yaga: the Ambiguous Mother and Witch of the Russian Folktale. Peter Lang.

Laparenok, Leonid. Prominent Russians: Vladimir I. Russiapedia (Get to know Russia better). Retrieved from https://russiapedia.rt.com/prominent-russians/history-and-mythology/vladimir-i/.

McCall, Gerrie. (2007.) Dragons: Fearsome Monsters from Myth and Fiction. New York, NY: Tangerine Press.

McCormick, Kylie. (2003, September 9.) Chudo Yudo. Dragons of Fame. Retrieved from ww.blackdrago.com/fame/chudoyudo.htm.

Niles, Doug. (2013.) Dragons: the Myths, Legends, & Lore. Avon, Massachusetts: Adams Media.

Rose, Carol. (2000.) Giants, Monsters, and Dragons: An Encyclopedia of Folklore, Legend, and Myth. New York, USA: W. W. Norton & Company, Inc.

Slaveykov, Racho. (2014.) Bulgarian Folk Traditions and Beliefs. Sofja: Asenevci.

Viasova, Eugenia. (2011, October 21.) Russian Dragon. Proper Russian. Retrieved from blog.properrussian.com/2011/10/russian-dragon.html.

Walsh, T. F. (2013, February 11.) Mythology Monday: Zmey, The Slavic Dragon. Retrieved from https://tfwalsh.wordpress.com/2013/02/11/mythology-monday-zmey-the-slavic-dragon/.

Warner, Elizabeth. (2002.) Russian Myths. Austin, TX: University of Texas Press.

Zmaj and the Dragon Lore of Slavic Mythology. (2015, January 5.) Ancient Origins: Reconstructing the Story of Humanity's Past. Retrieved from www.ancient-origins.net/myths-legends-europe/zmaj-and-dragon-lore-slavic-mythology-002984.

Zmey Gorynych. (2011, December 5.) Dragon's Corner: Exploring dragonkind around the world! Retrieved from https://dragonscorner.wordpress.com/2011/12/05/zmey-gorynych.

Dragon of Krakow

Background:

The first written mention of the Dragon of Krakow is in the Chronicles of Poland by Wincenty Kadlubek in the 13th century. But the story was likely told long before.

Myth:

The city of Krakow started as a small village on Wawel Hill, possibly back in the Stone Age. Many people travelled through the area. It was in a great position for trade from many directions. A king, known as Krak, developed the city as the center of his kingdom. But, it also became known that in a very deep cave in Wawel Hill, there was a large, green speckled egg. One day the egg cracked open and a dragon, Smok Wawelski emerged.

The young dragon was very hungry and began devouring sheep and cattle from the local farms, sometimes it is even said that it stole young women. The king, Krak, was overwhelmed with the misery that his people and kingdom were enduring.

Many brave men tried to kill the dragon but to no avail. Krak was desperate. Finally, a young shoemaker, Skuba Dratewka, came to the king with a plan. Skuba took a sheepskin and used his shoemaking skills to stitch the sheepskin to look like a living sheep, only Skuba filled it with sulfur from a local quarry. Then he took it to the mouth of the cave and waited to see what the dragon would do.

The dragon, Smok, came out and found the dead sheep full of sulfur and greedily devoured it within minutes. The sulfur started to burn inside his stomach. His stomach was on fire! So the dragon ran to the River Vistula and drank and drank and drank water until he exploded. The king offered the shoemaker, Skuba, a prize. Some say Skuba asked for the dragon's hide to make shoes, especially for the poor. The king gave Skuba a shop near the center of the city and he became famous and wealthy making exceptionally strong shoes and boots from the dragon's hide. Others say that Skuba asked for the hand of Krak's daughter in marriage and that Skuba became king when Krak died and built the first castle atop Wawel Hill.

Some stories say that Krak was actually the poor shoemaker who defeated the dragon, married a princess, and later became king. What is still told is how the city was saved from the dragon, and that it is the dragon's bones which hang today outside Wawel Cathedral.

Illustration:

The Dragon of Krakow, also known as Smok Wawelski, has forever been immortalized in science when Polish scientists named the carnivorous Triassic archosaur after it (*Smok wawelski*). This illustration is based on this real fossil found in southern Poland.

SOURCES:

Colavito, Jason. (2015, December 5.) The Fossil Dragon Bones of Poland's Wawel Cathedral. Jason Colavito. Retrieved from http://www.jasoncolavito.com/blog/the-fossil-dragon-bones-of-polands-wawel-cathedral.

Harrington, Jo. (2014, October 1.) Krakow and the Legend of the Wawel Dragon. Retrieved from www.wizzley.com/krakow-dragon.

Krakow info-Dragon Den. krakow-info.com. Retrieved from http://www.krakow-info.com/smocza.htm.

McCall, Gerrie. (2007.) Dragons: Fearsome Monsters from Myth and Fiction. New York, NY: Tangerine Press.

Monte, Richard. (2008.) The Dragon of Krakow. London, UK: Frances Lincoln.

Wawel. Wawel.krakow. Retrieved from www.wawel.krakow.pl/en/index.php?op-11.

Polish Legends, Myths and Stories. Anglik.net. Retrieved from http://www.anglik.net/polish_legends_dragon.htm.

The Bones of the Wawel Dragon. Atlas Obscure. Retrieved from www.atlasobscure.com/places/the-bones-of-the-wawel-dragon.

Puk

Background:

A puk, also called a puki, puuk,or pucky in some cultures, is the term for a small dragon spirit of German folklore.

Myth:

Puks are considered household dragons and are usually of a capricious nature. They are small, walk on four legs, and have a fiery tail. Some are said to take on the appearance of a housecat when walking on the ground. Some also have wings and can fly wherever they please.

Puks usually have a friendly demeanor, and will bring good fortune and gifts of jewels and treasure to their masters (although sometimes this happens at the expense of neighbors). They will guard the treasures of their home with great ferocity, but can be mischievous as well. If one refuses the gifts brought by a puk, the puk will take great offense and bring misfortune to the house in anger.

Illustration:

The small and mischievous Puk is drawn here with the likeness of the great crested newt (*Triturus cristatus*). These amphibians have fiery-colored underbellies and the males grow an impressive dorsal crest during breeding season.

SOURCES:

Bane, Theresa. (2015.) Encyclopedia of Beasts and Monsters in Myth, Legend, and Folklore. Jefferson, North Carolina, USA: McFarland & Company, Inc.

Rose, Carol. (2000.) Giants, Monsters, and Dragons: An Encyclopedia of Folklore, Legend, and Myth. New York, USA: W. W. Norton & Company, Inc.

Cultural Questions:

1. In Alkha, we see the sun and moon cycles as a prominent part of this myth. What other dragon myths include these concepts? Compare the cultural and ecological contexts.
2. Bolla is yet another dragon legend that focuses on water. This time in an area that is known to be prone to drought. Is Bolla an explanation for rain and floods?
3. Discuss the significance of Bolla becoming Kulshedra.
4. The numbers of 3, 9 and 12 recur frequently in Slavic myths. What is the significance of these numbers in the Slavic cultural systems?
5. The Chudo Yudo myth and its similar incarnations contain extensive descriptions of familial activities including revenge and brotherly support. Do the different family activities represent different societies or different lineages?
6. Two versions of the Krakow dragon legend exist. One, the hero is a prince, the other the hero is a shoemaker. What is the significance of the two very different heroes?
7. Local folk make noise to scare off Alkha by throwing rocks and making noise similar to the custom of scaring Nian. The gods came up with a different solution. Discuss the two solutions.
8. Is the Puk a dragon?

CHAPTER 11: WESTERN EUROPE

Background:

Western Europe is the home of the most common stereotypes of dragons: large, winged, 2-4 legs, fire breathing monsters of lore. Some of the myths that follow fit the stereotype; some do not. Some show definite influence of the arrival of Christianity. Others seem to exhibit influence from other directions.

Cuèlebre

Background:

In the northern area of Spain are the Cantabrian and Asturian regions with their own particular cultural heritage. This mountainous region on the north coast of Spain has often been the home of Asturian cultural autonomy. Neolithic influences came from the Atlantic and Celtic areas of the British Isles, and the stone described in this story can be found in similar Celtic tales. Christianity and Islam are more modern influences.

Myth:

The Cuèlebre is a very, very large serpent like dragon with bat wings and scales. Some say that the species is immortal yet others tell that they can be killed if you get them to eat bread full of pins or red hot stones. As they age, their scales become very, very thick, so thick as to be almost impenetrable, maybe that's why some say they are immortal. They are known to hoard precious jewels and metals. One myth claims that its spit can turn into a magical healing stone. The way the healing stone is collected changes from region to region in Asturias- some say you must collect the stone while the snake is alive to guarantee its potency. Other communities say it must be collected using a silver tray, and others insist on using smoking paper.

When they reach a great old age, the stories say that these serpents fly off to a paradise that exists beyond the sea, Mar Cuajada.

One story, told by many is of a vain young woman with long golden hair. She loved to comb her hair all day next to a spring inhabited by a strong willed nymph. Even though her mother and grandmother warned her that the nymph did not like her spring being disturbed, the girl would go and sit there and comb her hair. This meant that the girl ignored her duties and responsibilities to the family, yet still she sat and combed her long glorious golden hair.

The nymph kept watch on the girl, but couldn't do anything unless the girl actually disturbed the water of the spring. Finally, one day, one of the long golden hairs fell onto the water of the spring, disturbing the water. The nymph took the opportunity to give retribution for the girl's vanity and irresponsibility. The nymph cursed the girl into a dragon, in fact a Cuèlebre of gigantic proportions. Only a brave young man who could see the dragon as beautiful could free her. It is said that she still hides in a cave by the sea

.

Illustration:

Upset about her curse, this Cuèlebre hisses in anger. She is drawn using typical european wyvern design and the loose patterning of *Vipera aspis*, a venomous snake found in western europe.

SOURCES:

Álvarez Peña, Alberto Fundación Belenos. (2007.) Elementos de la Antigüedad Celta en la Tradición Oral Asturiana. Etnoarqueologia 13(53), 243.

Bane, Theresa. (2015.) Encyclopedia of Beasts and Monsters in Myth, Legend, and Folklore. Jefferson, North Carolina, USA: McFarland & Company, Inc.

Chisholm, Hugh, ed. (1911.) Cantabri. Encyclopædia Britannica, 5 (11th ed.). Cambridge University Press.

Pike, A.W.G. & Hoffmann, D.L. & García-Diez, Marcos & Pettitt, Paul & González, José Javier & De Balbin-Behrmann, Rodrigo & González-Sainz, C & de las Heras, C & Lasheras, J.A. & Montes, R & Zilhão, João. (2012). U-Series Dating of Paleolithic Art in 11 Caves in Spain. Science 336(6087), 1409-1413.

Rose, Carol. (2000.) Giants, Monsters, and Dragons: An Encyclopedia of Folklore, Legend, and Myth. New York, USA: W. W. Norton & Company, Inc.

The Cuelebre. Draconika. Retrieved from http://www.draconika.com/legends/cuelebre.php.

The Cuelebre. (2015, August 31.) Valley of Dragons. Retrieved from http://valleyofdragons.com/the-cuelebre/.

Graoully Dragon

Background:

Many stories are told of Saints and dragons. Some say this is because the Bible says that saints can confront snakes and not be harmed.

The town of Metz is in northeastern France. Christianity was first being taught there to the folk of Europe by followers of the first disciples. Some say it was St. Peter, some say it was St. Pierre, who sent (later to be known as) St. Clement with two of his own disciples, Celestius and Felix to northeastern France to convert pagans to Christianity. Upon reaching the town of Metz, St. Clement is said to have saved a stag from being killed in the hunt, but then he did an even more amazing thing.

Carved depictions of the Metz dragon as far back as the 1100s exist, and have been argued to be based on fossils found in the area. Some scientists believe that depictions of the Metz or Graoully dragon correspond to an extinct ichthyosaur whose bones would have been found in the area. Alternately, the number of snakes in the story could potentially represent recurrent plagues.

Myth:

The town of Metz was being overwhelmed by a large number of snakes who lived in the local Roman amphitheater. The vapors from these snakes were horrendous, so foul that they kept the folk of Metz confined to their local township. The greatest of these snakes was known as Graoully, a very large, scary dragon-like snake.

St. Clement declared that he could and would free the town of the snakes if they swore to convert to Christianity. St. Clement entered the Roman amphitheater with only his sacred clothes and a cross. The snakes attempted to close in and kill the Saint, but St. Clemens made the sign of the cross and the snakes were tamed. (In another version, he throws his stole around the dragon and it is tamed, walking behind him like a dog, and all snakes, lizards, and other such creatures are along tamed with it.) Then St. Clement led the Graoully and all of the other snakes to the River Seille. There he told them to disappear and never come near people again. They did, and the town converted to Christianity.

Illustration:

The Graoully Dragon bends away from St. Clement. He is drawn with bat wings and the pattern of *Natrix natrix*, the grass snake, native to France and non-venomous.

SOURCES:

Bellard, A. (1968.) Le Graoully de Metz à la lumière de la paléontologie. Mémoires de l'Académie nationale de Metz, 139-146.

Delehaye, Hippolyte. (1955.) The Legends of the Saints. Dublin: Four Courts Press.

Ingersoll, Ernest. (1928, 2014.) Dragons and Dragon Lore. New York: Cosimo Classics.

Legend of the Graoully Dragon and Saint Clement of Metz. (2012, December 12.) Green Mountain College: The Rope Swing. Retrieved from https://theropeswingblog.wordpress.com/2012/12/29/legend-of-the-graoully-dragon-and-

saint-clement-of-metz/.

Legends. Visit Me TZ: Office de Tourisme Metz Métropole. Retrieved from http://www.tourisme-metz.com/en/legends.

McCullogh, Joseph. (2013.) From Beowulf to St. George. Bloomsbury Publications.

Nevins, Scott. (2015, April 26.) The 'Dragon-Slayer' Saints of the Eastern Orthodox Church. Scott Nevins Memorial. Retrieved from https://scottnevinssuicide.wordpress.com/2015/04/26/the-dragon-slayer-saints-of-the-eastern-orthodox-church/.

Privat, Jean-Marie, dir. (2006.) Dragons entre sciences et fiction. Paris, CNRS Éd.

Lambton Worm

Background:

In the northeastern part of England, some say the story took place in Northumbria, some say in Durham, there is a tale still told of the Lambton Worm, a very large worm-like serpent. Over time, both Saxons and Christians fought dragons/worms in this area.

The number nine is very significant in Chinese dragon mythology: nine sons of the dragon, etc. Sea lampreys are common in this area. They have a row of seven gill holes. If you add in the eye and single nostril at the top of the head, one could easily see where the image of a nine-eyed eel comes from.

Myth:

This myth is set during the times of the many Crusades. The myth starts with a young man, the heir to the Lambton family, who went fishing in the Wear River on a Sunday morning when he ought to have been in church. He didn't catch any fish. All he caught was a worm. Some say the worm was two feet long, some say it was the size of a finger, whatever the truth, John Lambton thought the worm was useless even for eating and threw the worm down the local well.

As Lambton gained maturity, he decided to atone for his disrespectful behavior and went on Crusade for seven years. However, while he was gone, the worm grew and grew and grew. It grew to the size that some say it could wrap itself around a hill nine times.It had nine gill slits on both sides of its neck. It was a ravenous creature, devouring cattle, sheep and whatever it wanted, even small children, especially at night. The local folk were, understandably, terrified. No one knew how to kill such a creature.

Then John Lambton returned home to his elderly father and ancestral home. He discovered that the worm that he had thrown into the local well had grown into a very scary monster. He sought advice, some say from an old man (a family servant), some say the wise woman of Brugeford or a witch. John was advised to create a suit of armor with sharp blades all over the outside then entice the worm to fight him. John created this unusual suit and went to battle the worm. The worm wrapped himself around John, cutting itself into many pieces and then died.

The advice had come with one stipulation: after killing the worm, John must kill the first living thing he encountered. John set it up that his dog would be released at the appropriate moment and that would be the first living thing that John encountered. However it happened, it was not the dog that John first met, but his father. He just couldn't kill his father. The result was that for the next nine generations, no Lambton heir died peacefully in their bed until Henry Lambton, M.P., in 1761.

Illustration:

John Lambton confronts the terrible Lambton Worm. The myth describes the worm with 'nine gill slits,' which brings to mind the sea lamprey (*Petromyzon marinus*), which has seven gill holes on each side of its body, two eyes, and one nostril—for this reason it is sometimes called the "nine-eyed eel." It is this fish which inspires the design of this dragon.

SOURCES:

Allen, Judy and Jeanne Griffths. (1979.) The Book of the Dragon. New Jersey: Garnstone Press.

Bane, Theresa. (2015.) Encyclopedia of Beasts and Monsters in Myth, Legend, and Folklore. Jefferson, North Carolina, USA: McFarland & Company, Inc.

Dekirk, Ash. (2006.) Dragonlore: From the Archives of the Grey School of Wizardry. New Jersey: Career Press, Inc.

Hargreaves, Joyce. (2009.) A Little History of Dragons. New York: Walker & Company.

Ingersoll, Ernest. (1928, 2014.) Dragons and Dragon Lore. New York: Cosimo Classics.

Jacobs, Joseph. (1894.) More English Folktales. London: David Nutt and Co.

McCall, Gerrie and Regan, Lisa. (2011.) Monsters and Myths: Dragons and Serpents. New York, NY: Gareth Stevens Publishing.

Niles, Doug. (2013.) Dragons: the Myths, Legends, & Lore. Avon, Massachusetts: Adams Media.

Rose, Carol. (2000.) Giants, Monsters, and Dragons: An Encyclopedia of Folklore, Legend, and Myth. New York, USA: W. W. Norton & Company, Inc.

Sea Lamprey. British Sea Fishing. Retrieved from http://britishseafshing.co.uk/sea-lamprey/.

Shuker, Karl. (1994.) Dragons: A Natural History. New York: Barnes & Noble Books.

Longwitton Dragon

Background:

Throughout Britain, there are many stories of healing wells of water. They are often referred to as Treacle Wells. The meaning of treacle comes from ancient sources which thought that some wells had water which contained medicinal elements.

Many were renamed after Saints when Christianity came to dominate the land.

Myth:

There are three healing wells in the woods near the town of Longwitton in Northumberland, England. The three treacle wells near Longwitton were and still are known wide and far for their healing powers. Many folk including sick children, tired shepherds, weary travellers and many more came to these wells to drink of the healing water and be restored.

One day a plowman (someone who uses a plow, like a farmer) went to the wells for a restorative drink, but when he arrived at the well, there was a very large, green dragon with poisonous glands along its body, an extremely long tail and a head with a crest and teeth the size of knives. From its mouth, from between the teeth, green venom dribbled down, its origins in the venomous sacs below the dragon's jaws.

The dragon was drinking the water from the well. Then the dragon disappeared. The plowman could still feel the dragon nearby, so he ran really fast back to the village.

From that day on, only the dragon drank from the wells. It coveted the water from the wells like some dragons are said to protect their golden treasures. Finally a young knight came, one who wanted to fight for glory and declared that he would vanquish the dragon. The townsfolk hoped he could.

The young knight was none other than Sir Guy of Warwick. Sir Guy was in possession of a magical ointment which he had brought back from his many travels. The ointment when applied to his eyes would allow Sir Guy to see the dragon. Sir Guy went to the wells and fought the dragon for a full day, but every time he wounded the dragon, the wounds, mortal or not, would miraculously heal themselves. At the end of the first day, Sir Guy dragged himself back into the town, defeated.

But Guy would not quit. He went to the wells on the second and again put up a valiant fight, only to, again, be defeated as the dragon's severe wounds healed themselves.

On the third day, Sir Guy decided to observe the dragon's movements as he parried with the beast. He noticed that the dragon always kept the tip of its tail in one of the wells. So, thinking fast, Sir Guy and his noble steed started to fight but in a way that slowly drew the dragon away from the wells. Sir Guy and his horse pretended to be dying and the dragon decided to get close for the kill. As soon as the dragon was far enough away from the wells, Sir Guy on his horse rode quickly to cut off the dragon from the healing waters. Then Sir Guy was able to slay the dragon. Some say the dragon disappeared into a pool of slime, quickly absorbed by the healing waters.

The townsfolk rejoiced and the three treacle wells of Longwitton returned to their destiny, to heal all who came to drink from the wells.

Illustration:

Keeping its tail in the well-waters, the Longwitton Dragon confronts Sir Guy. Drawn in the typical style of medieval european dragons, the Longwitton Dragon also uses the distinctive pattern of *Vipera berus*, Britain's only venomous snake.

SOURCES:

Longwitton (Northumberland). Retrieved from http:www.wyrm.org.uk/ukdracs/longwitton.html.

The Three Treacle Wells of Longwitton. (2011, March 18.) Leaves in the Lonnen. Retrieved from https://fettlereetly.wordpress.com/2011/03/18/the-three-treacle-wells-of-longwitton/.

Varner, Gary R. (2009.) Sacred Wells: A Study in the History, Meaning and Mythology of Holy Wells. Algora Publishing.

Yolen, Jane. (2014.) Favorite Folktales from Around the World. Knopf Doubleday Publishing Group.

Nant Gwynant: The Red Dragon of Wales

Background:

Even before Roman colonization of the British Isles, the Indigenous communities in the British Isles were confronted with Saxon invasions. After Roman colonization, the assaults by the Saxons resumed. Out of the resistance by the Indigenous people of the British Isles come two myths which clearly symbolize the power struggle.

Myth:

The first legend told by the Indigenous peoples there is this. Beli the Great was the father of four sons: Lludd, Caswawallawn, Nyniaw and Llefelys. Lludd inherited the kingdom of Britain when Beli the Great passed on. Lludd was known for rebuilding the walls of London (Caer Lludd) with towers a plenty. Lludd was a "good warrior and generous and liberal in giving meat and drink to all who sought them", important traits for a great leader (Jones, p. 75). Lludd loved his younger brother, Llefelys, who was well known as a wise counselor.

Lludd ruled Britain and Llefelys ruled France. Each ruled well, but there came a time when three plagues befell Britain. The first was the people called "Coraneid". The second was a scream heard everywhere in Britain on Mayeve. The scream tore through the hearts of all. Warriors were weakened. Women lost their unborn babes. Young women and men lost all of their senses and all of the earth and its creatures and greenery became barren. The third was the disappearance of a year's worth of food and drink in one night.

King Lludd could not perceive how to defeat these plagues. He asked the advice of all of his nobles, who came together and advised Lludd to go to his brother, Llefelys, King of France and seek his counsel. Llefelys was a very wise person. He gave Lludd the information he needed to defeat all three plagues.

The first plague was that the Coraneid people were able to hear any word spoken out loud throughout the whole island. Llefelys ordered that a very long, bronze horn be made where only Llefelys and Lludd could hear what they said to each other. Llefelys advised that Lludd take some very specific insects, crush them into water, and bring that to his Homeland. There Lludd should invite the Coraneid and his own folk to a feast to make peace. Lludd should spread the special water over all. It only killed the Coraneid.

It is the second plague that is of concern here. Llefelys said it was the scream of a dragon, the red dragon of Lludd's homeland. It was fighting a white dragon, the dragon of people trying to conquer Lludd's people (some say it was the Saxons).

Llefelys's instructions were for Lludd to return to Britain and measure the island's length and breadth to find the exact center. The center was Oxford. In that place, as instructed by Llefelys, Lludd had a pit dug. A tub of the best mead was placed in the pit. A silk covering was placed over the tub. Lludd kept watch himself and saw the dragons fighting. After much fighting, the tired dragons sat on top of the silk covering, which sank with them to the bottom. Then the dragons drank all of the mead and fell asleep. Lludd immediately folded the silk covering over the dragons. Lludd then took the dragons to the safest place in the kingdom that he knew, Dinas Emreis (formerly known as Dinas Ffaron Dandde) and buried them in a stone coffer. That ended the scream.

By the way, the third plague turned out to be a giant of a man who came each night and stole away with all of the food and drink. Lludd defeated him in combat and the land was peaceful for many years.

Many centuries later, some say as many as five, after the Romans had left the island, King Vortigern, was battling both the Picts and some of the Saxon tribes. King Vortigern made an alliance with some of the Saxon tribes in return for land. These tribes helped Vortigern fight the Picts and other Saxons, but after a while, these Saxons became more and more greedy, wanting more and more land. Vortigern met with his twelve wisest men who said that Vortigern should build as strong a castle (with a city inside) as possible.

The place chosen was Dinas Emreis. Stone masons, carpenters, builders of all kinds gathered there with all of the necessary building materials. However, the first night the building materials disappeared. Again, stones and all were gathered and on the second night, they, too, disappeared, as if the earth had just swallowed them up. A third night, and again, all disappeared. Vortigern summoned his Wise Men again and this time they said that Vortigern "must find a child born without a father" (Thomas 1907, Roger Lnacclyn Gree, ed.). The boy must be killed and his blood spread where the castle is to be built.

So, Vortigern sent messengers everywhere in Britain. Finally, such a boy was found and the boy and his mother were brought to Vortigern's court. The boy was Merlin whose name was also Emrys. Merlin informed Vortigern that the Wise Men were ill-informed, that they did not know what lay under the intended foundation. Merlin asked Vortigern to have some men dig into the ground. The men came upon a deep pool where Merlin pointed out there were two stone coffers, each with a dragon, one red, Vortigern's dragon and one white, the Saxon's dragon. When the coffers were opened, the dragons arose and started fighting. The White Dragon of the Saxons drove the Red Dragon to the edge and almost out, several times.

Finally, the Red Dragon rallied and drove the White Dragon out. And while Arthur ruled the land, the Saxons were kept at bay. But, after Arthur passed, the Saxons gained dominance of Britain. However, Merlin predicted that eventually, over much time, they would be driven out. Vortigern did not build a castle at Dinas Emrys. That's where Merlin lived.

Illustration:

The Red Dragon of Wales, upper dragon, engages the White Dragon of the Saxons in combat, lower dragon. The Welsh flag helped inspire the design of the red dragon.

SOURCES:

Ashley, Mike, ed. (2002.) Arthurian Legends. New Jersey: Castle Books.

Green, Roger Lancelyn. (1973.) A Book of Dragons. New York: Penguin Books.

Jones, Gwyn and Thomas Jones. (1949, 1991.) The Mabinogion. London: Everyman.

Walton, Evangeline. (2004.) The Mabinogion Tetralogy. New York: The Overlook Press.

Lyminster Knucker

Background:

In southern England, there is a village Lyminster located in West Sussex. A knucker is a dragon. Some say the word comes from a Saxon word, "nicor" which means "water monster". It is mentioned in Beowulf. Knuckers are exceptionally large, long dragons with a large head, mouth and long slithery tail. They have a loud roar, enough to shake the earth and a venomous bite.

To this day, St. Mary Magdalene Church in Lyminster still displays a monument called the Slayer's Slab, commemorating the defeat of the knucker. The Sussex area had many dragon myths, with the neighboring areas of Bignor Hill and St. Leonard's Forest also having local dragon stories.

Myth:

The legends told vary some but here is the general tale. In the village of Lyminster there is a Knucker's Hole, a virtually bottomless pit which was home to a knucker, a giant extremely hungry knuckler. It is said that every night it came forth to eat whatever meat, four legged or two legged that it could find. Some say it preferred young women, preferred them so much that soon there were few left. Whatever the truth to the dragon's dietary preferences, there came a day when the King of Sussex and the Mayor of Lyminster offered rewards. The King offered the hand of his daughter in marriage and the Mayor offered great riches.

Several knights came to town to slay the dragon. None were successful. Finally, a young man named Jim (Pulk or Puttock, depending on the source) came up with a plan. The plan required the help of the local bakers. A very large poisonous pie was baked and loaded on a cart pulled by a horse, and Jim drove it to Knucker's Hole to lure the dragon to eat it.

Jim was unable to unload the cart before dark so he hid nearby and awaited the dragon's appearance. Finally, the dragon came out of the Hole and smelled the luscious pie. He ate the pie, horse and cart all. The poison took a while to take full effect, so Jim followed the dragon until it fell down, exhausted, close to death. Then Jim came out of hiding and with his axe, cut off the head of the dragon and brought the head back to town as proof of the dragon's death.

Since Jim was not a noble person, the King did not give him the hand of his daughter in marriage. The Mayor did bestow great riches, enough for Jim to buy a beautiful house and estate and live very, very well for the rest of his days. When he died, he was buried under a stone, engraved, "Slayer's Stone" which exists to this day in the nearby Saint Mary Magdalene's Church. Some stories say that this Jim was named Jim Puttock, and a competing version about a Jim Pulk says that he did bake a poisoned pie, but ingested a bit of the poison himself and tragically died after defeating the dragon.

Illustration:

Ever gluttonous, the Lyminster Knucker reaches for a delicious pie. A cross between a typical european dragon and the common toad (*Bufo bufo*), this is a very fat dragon indeed.

SOURCES:

Bullion, Alan. (1978.) English Dragon Legends Number 3. Journal of Geomancy 2(3).

Bane, Theresa. (2015.) Encyclopedia of Beasts and Monsters in Myth, Legend, and Folklore. Jefferson, North Carolina, USA: McFarland & Company, Inc.

Dekirk, Ash. (2006.) Dragonlore: From the Archives of the Grey School of Wizardry. New Jersey: Career Press, Inc.

Dragons & Serpents in Sussex. Sussexarch.org.uk. Retrieved from http://www.sussexarch.org.uk/saaf/dragon.html.

Hargreaves, Joyce. (2009.) A Little History of Dragons. New York: Walker & Company.

McCall, Gerrie. (2007.) Dragons: Fearsome Monsters from Myth and Fiction. New York, NY: Tangerine Press.

Niles, Doug. (2013.) Dragons: the Myths, Legends, & Lore. Avon, Massachusetts: Adams Media.

O'Leary, Michael. (2013). Sussex Folk Tales. The History Press.

Tarasque

Background:

In the south of France, in the area of Provence, there is an ancient myth of a gargantuan monster. It was thought to be the progeny of Leviathan and another creature. It had a large body with a head like a lion, teeth as sharp as swords, a horse's mane, sharp scales on the sides of its body and a turtle like shell on its back. It had 6 legs with bear claws for feet. And then, of course, it had a long serpent like tail. It was ferocious with a voracious appetite. The local folk were in abject fear of it with no idea how to defeat the ever hungry dragon.

Myth:

One day, a ship arrived; some say it was 48 A.D./C.E. On the ship were Lazarus, Mary Magdalene and their sister, Martha. Saint Martha she became and one of her saintly deeds was to defeat the dragon of Tarasque.

When the three got off of the ship, they were told of the vicious creature and it was Martha, a strong willed woman, who went in search of the monster. She found it quickly and came upon it without the monster noticing. It was busily devouring its recent meal, a man. When the creature saw her, she picked up two sticks and made the sign of the cross. She also threw holy water on the beast. It immediately became docile and she led it back to the town.

The townspeople were very afraid and attacked and killed the dragon. It was under St. Martha's control and did not fight back. St. Martha was not happy about the attack. She spent time teaching the words of Jesus Christ about peace and forgiveness.The townspeople converted to Christianity and felt remorse for their behavior.

To this day, the town of Tarasque celebrates St. Martha's saintly deed with a town wide feast and parade.

Illustration:

Many sculptures have been made in the image of the Tarasque. This drawing takes inspiration from these, as well as lion, horse, and leatherback sea turtle (*Dermochelys coriacea*) parts. While they don't have six legs, Dermochelys is the largest of all living turtles, and can weigh over 1000 pounds.

SOURCES:

Bane, Theresa. (2015.) Encyclopedia of Beasts and Monsters in Myth, Legend, and Folklore. Jefferson, North Carolina, USA: McFarland & Company, Inc.

Ingersoll, Ernest. (1928, 2014.) Dragons and Dragon Lore. New York: Cosimo Classics.

Meurger, Michel and Claude Gagnon. (1998.) Lake Monster Traditions: A Cross-Cultural Analysis. London: Fortean Times.

Niles, Doug. (2013.) Dragons: the Myths, Legends, & Lore. Avon, Massachusetts: Adams Media.

Of cooks, pirates, and dragons: Saint Martha. (2011, July 29.) A Nun's Life Ministry. Retrieved from https://anunslife.org/blog/nun-talk/of-cooks-pirates-and-dragons-saint-martha.

Rose, Carol. (2000.) Giants, Monsters, and Dragons: An Encyclopedia of Folklore, Legend, and Myth. New York, USA: W. W. Norton & Company, Inc.

Shuker, Karl. (1994.) Dragons: A Natural History. New York: Barnes & Noble Books.

Simpson, Jacqueline. Tarasque. Retrieved from folklore-society.com/miscellany/tarasque.

Saffron Walden Basilisk

Background:

The Saffron Walden Basilisk is commonly referred to as "basilisk" coming from the Greek word, "basiliscus" meaning "little king". Stories start as early as in Pliny the Elder's (Roman) writings on natural history and continue through the medieval times but morphed during medieval times into the cockatrice.

Myth:

Legends say that the basilisk is actually a small serpentine creature, ranging from six inches to two feet, but size doesn't matter with this most venomous beast. Its breath, its touch, its saliva, its scathing look could kill. Some say it could spit into the air to kill birds and that the touch of its body killed all living things, plant and animal alike. Some say that there used to be so many of these lethal beings that their touch on the earth created the deserts of the Middle East and Africa.

On its head was a crown or crest of three protuberances which resembled a crown and lent to the origin of its name as a little king. This may have started as a white mark, that was elaborated in later stories to be a gold mark or crown. And it could stand on two legs.

The basilisk could be killed in three ways: by a weasel (a natural enemy with a venomous bite), the crowing of a cock (the sound basically drove the basilisk crazy unto death), and its own reflection. As its look could kill humans, if it saw its own image in a mirror, it could kill itself.

Over time, especially during the medieval ages, the basilisk began to change into what is now called a cockatrice, an even more fantastical creature, hatched from a seven year old rooster's egg. In addition, the cockatrice had the legs, neck and head of a rooster.

Illustration:

With is petrifying gaze and distinct 'crown,' stories of the Basilisk almost surely arose from tales of the spitting cobras (genus: *Naja*) of Africa and Asia. These snakes have specialized fangs which can shoot venom out at aggressors' eyes, potentially causing blindness.

SOURCES:

Bane, Theresa. (2015.) Encyclopedia of Beasts and Monsters in Myth, Legend, and Folklore. Jefferson, North Carolina, USA: McFarland & Company, Inc.

Hargreaves, Joyce. (2009.) A Little History of Dragons. New York: Walker & Company.

Jobes, Gertrude. (1962.) Dictionary of Mythology Folklore and Symbols. New York, NY: The Scarecrow Press, Inc.

Rose, Carol. (2000.) Giants, Monsters, and Dragons: An Encyclopedia of Folklore, Legend, and Myth. New York, USA: W. W. Norton & Company, Inc.

Shuker, Karl. (1994.) Dragons: A Natural History. New York: Barnes & Noble Books.

Mordiford Wyvern

Background:

Many myths are told of wyverns, two legged, two winged, serpent-like creatures. They are often like small versions of dragons, and some compare them to the ancient flying reptile group, the pterosaurs. The word is said to have several sources; Latin vipera meaning viper or asp, also Saxon wivere meaning serpent and in Anglo-Saxon it means a lindworm, a legless dragon. It was originally seen as a life bringer, but after the acceptance of Christianity, it became the vicious predator in the following story (and the many versions of this story).

Myth:

Whatever the derivation of the word, the stories focus on the town of Mordiford in the southwest of England. Many, many years ago, before stories were written down, the folk of Mordiford told of a wyvern who threatened the town and all of its folk and their herds, killing whenever it felt the need for a good meal. It developed an especial liking of human flesh. The story was painted on the wall of the local church. It was repainted many times over the centuries, sometimes having a different number of wings and legs, but the dominant picture was of a two legged, two winged serpent with a pointy tail.

The stories say that the wyvern was killed as the wyvern feasted on a dead ox that had been washed up along the river shoreline. In its engorged exhaustion, the villagers were able to creep up and kill it.

Other versions say it was killed either by a convicted criminal named Garstone who offered to kill it for a pardon or a local young nobleman from the Garstone family who built a barrel covered with all manner of knives and pointy instruments of destruction. Some say it was a cider barrel, the wyvern's favorite drink. Whether it was the odor of cider or the smell of its favorite meal, human flesh, the wyvern approached the barrel and tried to consume it. This was a fatal mistake as the sharp points all over the barrel soon wounded the wyvern until it was close to death. Some say that the wounds were mortal. Others say that the man inside the barrel then shot him with a pistol, but that would be a more recent ending.

The painting was on the church until 1812 when the church underwent restoration. Some say that the vicar at the time related the dragon to the devil and forbade painting it on the church. Others say people just didn't get around to re-painting the wyvern.

In 1864, a new version of the Mordiford wyvern is told. This version adds the element of a young girl, Maud, who wanted a pet of her own. Maud liked to visit the local forest and, one day, found a small, two winged creature stumbling around and brought it home as her new pet. Her parents quickly realized that the creature was a wyvern and absolutely forbade keeping it in the home. Maud was very sad and decided to raise the wyvern in secret in a special hiding place in the forest. She fed the wyvern a steady diet of milk. It grew rapidly. After a while, milk just wasn't enough. It wanted flesh, sheep, then a cow, then a farmer who tried to chase the wyvern away from his herd. The wyvern developed a taste for human flesh as the stories say that most wyverns do.

Maud pleaded with the wyvern to stop its murdering ways, but the wyvern did not, although it always remained a friend to Maud. Finally, the townsfolk sought help and found a man, either the criminal or the nobleman, to come to kill the wyvern. Then the story continues with the barrel with sharp points all over and the death of the wyvern.

Some versions claim that with its dying breath, the wyvern killed its killer. Whether the man survived or not, the story claims that Maud, overcome with grief for the wyvern became absolutely hysterical and inconsolable.

Illustration:

Maud sleeps with her dear pet, the Mordiford Wyvern. Grown large from many meals of milk and meat, this dragon is loosely based on the typical descriptions of european wyverns at this time.

SOURCES:

Allen, Judy and Jeanne Griffiths. (1979.) The Book of the Dragon. New Jersey: Garnstone Press.

Bane, Theresa. (2015.) Encyclopedia of Beasts and Monsters in Myth, Legend, and Folklore. Jefferson, North Carolina, USA: McFarland & Company, Inc.

Freeman, Richard. Mordiford Wyvern. Mysteriousbritain.co.uk. Retrieved from http://www.mysteriousbritain.co.uk/england/hereford-and-worcestershire/legends/mordiford-wyvern.html.

Hargreaves, Joyce. (2009.) A Little History of Dragons. New York: Walker & Company.

Origins of Maud and the Dragon. (2006.) Myths and Legends. Retrieved from http://myths.e2bn.org/mythsandlegends/origins1353-maud-and-the-dragon.html#top.

Rose, Carol. (2000.) Giants, Monsters, and Dragons: An Encyclopedia of Folklore, Legend, and Myth. New York, USA: W. W. Norton & Company, Inc.

Shuker, Karl. (1994.) Dragons: A Natural History. New York: Barnes & Noble Books.

Buccoleon

Background:

In Ghent, Belgium, there is a belfry that overlooks the city with a golden dragon atop. The tower's construction was begun in 1313. It was completed in 1380. Some say that the golden dragon weathervane atop the belfry was made in Norway and gifted to Belgium.

Some tell another story.

Myth:

In the city of Aleppo, at the time of the many Crusades, lived an old dragon with brown scales. His scales were brown because Buccoleon was a good dragon and wept over what seemed like never ending fighting between the Saracens and the Belgian crusaders.

Buccoleon's tears were incredibly fertile, where they fell, flowers grew. The flowers were of many magnificent colors. They were called the Turk's Turban.

Finally, the wars were over and Taff, a man who loved flowers dearly, decided to bring seeds from the Turk's Turban back to his Flemish garden. The result was many, many beautiful flowers, now called tulips. Taff's tulips became famous. He made a handsome profit. The fame of the tulips spread far and wide, even reaching Aleppo.

When the wars stopped, Buccoleon's tears dried up and his scales changed to gold. This had been predicted and so it came to pass. But Buccoleon missed his beautiful flowers and when he heard of the tulips in Belgium, he decided to go and see them for himself. Along his long journey, he received directions from a marsh bird, a cloud, the Danube river, and the wind. Finally, as he flew along with the wind, he came over Taff's garden and saw the beauty of the field of many colored tulips.

Suddenly, an arrow came flying through the air and killed Buccoleon. Who would kill this good dragon? Many say it was goldsmiths and a baker, jealous of Taff's success and when they saw the flying gold of Buccoleon, they sought the golden scales for themselves. However, in Buccoleon's honor, the townspeople voted to place him atop the belfry, so all could see and honor him forever.

Illustration:

A flower grown from where Buccoleon's tears were shed. He is also drawn from the typical four-legged, leathery-winged descriptors of western europe.

SOURCES:

Dunton, Larkin. (1896). The World and Its People. Silver, Burdett.

Dragon of Ghent. Sacred Texts. Retrieved from www.sacredtexts.com/etc/tl/tl08.htm.

Griffs, William Elliott. (1919.) Belgian Fairy Tales. Thomas Y. Crowell Co.

Gulden Drak. Brouwerij Van Steenberge. Retrieved from www.vansteenberge.com/en/our-beer/gulden-draak/gulden-draak.

Lane, Bertha Palmer. (2009.) Tower Legends. Abela Pub. Ltd.

Drac

Background:

A myth is often told of dragons capable of luring humans and making them do whatever the dragons want. One well known tale centers on the Rhone river in the south of France where there is still a town known as Beaucarie.

As early as the 12th century, folk were telling stories of winged dragons who lived in the River Rhone, down deep below the surface. They were invisible to humans which made it easy for the dracs to capture unsuspecting folk floating on the Rhone in boats. If a person put their hand in the water, some say to grasp something shiny that the dracs made to appear, then the dracs would grab the person by the arm or hand and drag them to a watery depth where they would be devoured.

Myth:

One time, the captured woman was not devoured. She was a young mother still nursing her young child. The drac stole her away from her family and village to nurse his own son in his underwater cave. The woman fed the young drac her mother's milk for seven years. She forgot all about her life and her family. Each night she had to anoint the baby drac with magical ointment which would make him invisible to human eyes. Some say she mistakenly put the ointment in one of her own eyes. Others say she was supposed to anoint the baby drac with a special oil that made it visible to humans so she could see it to take care of it and one night forgot to wash her hands and got some of the ointment into one of her eyes. However, it happened, when the drac returned the woman, seven years later to her family, she had the ability to see the drac with one eye.

Most of the stories say that the woman no longer remembered her years with the drac. She returned to her husband and son, now seven years older and resumed her life.

Until one day, as she went to market to sell her flowers, she saw the drac looking over the townsfolk for his next meal. All of the memories came back and she started to rant and weep about what had happened. The drac, sensing danger to himself, ripped out the eye that could see him. The woman still raged on,but most folk assumed she was mad and she became a piteous and ragged soul, forever babbling on about invisible dragons.

Some say that no one believed her, but children still disappeared. Others say that a vast army was sent to find drac, but none ever found him.

Illustration:

Drac approaches an unsuspecting boat from below. This dragon is inspired by the Wel's catfish (*Silurus glanis*), which can grow over 7 feet long and has a diverse diet of everything from worms to ducks.

SOURCES:

Freeman, Richard. (2006.) The Drac. Myths. Retrieved from http://myths.e2bn.org/mythsandlegends/userstory8588-the-drac.html#top.

Meurger, Michel and Claude Gagnon. (1998.) Lake Monster Traditions: A Cross-Cultural Analysis. London: Fortean Times.

Rose, Carol. (2000.) Giants, Monsters, and Dragons: An Encyclopedia of Folklore, Legend, and Myth. New York, USA: W. W. Norton & Company, Inc.

Shuker, Karl. (1994.) Dragons: A Natural History. New York: Barnes & Noble Books.

Sonofskankware. (2012, August 20.) Drac. Mythical Creatures Guide. Retrieved from http://www.mythicalcreaturesguide.com/page/Drac.

Vandale, James. (1999, March 11.) The Drac: French Tales of Dragons and Demons. Reptilian Agenda. Retrieved from http://www.reptilianagenda.com/myth/m110399b.shtml.

Marraco

Background:

In the Basque language, the word, marraco, means dragon - in this case, yet another large green serpent. Not much of Basque mythology survived the introduction and domination of the Christian religion. However, in Catalan mythology, a culture just south of the Basque region, there is a tale of a marraco.

Myth:

The marraco in this myth is known to have such a wide mouth that it can eat human beings whole.

For the pre-Roman culture of the Ilergetes, or Ileida, who resided in the northern area of Spain, the marraco was the symbol of their deity, even thought to be the spiritual father of their most important political leader. This culture is thought to go as far back as the fifth century B.C./B.C.E.

While the Ilergetes culture may not still be around, ever since the Middle Ages, Catholic processions in this area include very, very large dragons without wings and large tusks. Sometimes it blows smoke through its nostrils. It's a major local attraction for tourists and for all.

A recent discovery in northern Africa claims that there is the depiction of a dragon that covers the desert area there and relate it to the Marraco dragon.

Illustration:

Spain is home to a great deal of dinosaur fossils, some of which belong to the armored dinosaur family Nodosauridae. Marraco is illustrated here using images of those fossils.

SOURCES:

Curcó i Pueyo, Jordi. (1996.) “Lo Marraco” i els gegants de Lleida i comarques. Catalonia, Spain: Ajuntament de Lleida.

Curcó i Pueyo, Jordi. (1987.) Els Gegants, Capgrossos i Lo Marraco de Lleida. Collecció La Banqueta: Lleida.

Hydra

Background:

In Greek mythology there are stories of the hydra. Most stories are of the Hydra of Lerna, allegedly the offspring of Typhon and Echidna, a couple known to have spawned more than one monster. But it is the Hydra of Lerna, defeated at the hands of Heracles, that so many stories are about.

Myth:

The Hydra of Lerna was a large monstrous serpent with at least nine heads (some stories say up to one hundred heads). The central head was thought to be immortal. Its breath was said to also be venomous - one whiff and you'd be dead. And its blood could also kill you with a mere drop. It lived near a lake, the lake of Lerna near Argolis in Greece. It had a voracious appetite, regularly scouring the fields of the land and devouring all of the flesh that it could find. Needless to say, this made the land a very desolate place.

The Hydra lived in a cave in the area, said to be a portal to the Otherworld. It came out whenever it was hungry, which was all too often for the local folk. Heracles (or Hercules as some refer to him) was sent to dispatch the Hydra by Eurystheus, as part of twelve heroic deeds that Heracles needed to complete. Some say this was to achieve immortality, others say it was to atone for killing his own family in a fit of madness inflicted upon him by Hera.

Killing the Hydra was no easy task and required the help of Heracles's nephew, Iolaus. First, Heracles shot flaming arrows into the Hydra's cave, to draw it out. Then Heracles began to hack off the many heads. However, as Heracles cut off a head, another grew in its place (some stories say two grew back for each one). The battle was wearying and finally Heracles paused and conferred with his nephew. The new attack included a flaming torch, and as Heracle cut off a head, Iolaus would cauterize the stump, preventing any head from growing back.

After much hacking and beheading, finally there was only the one immortal head left. This head was cut off, but it stayed alive, hissing at Heracles as he dropped it into a deep hole and placed a boulder on top so that it could never again emerge and harass the area. Heracles then dipped his arrows into the poisonous blood of the hydra, so that any wound inflicted with them would be deadly.

Illustration:

Here, the Hydra emerges from between the mountains with a countless number of heads. The myth of the Hydra surely draws some influence from the polycephalic condition, where rarely newborn animals will have two heads or two faces. While anecdotally polycephally is more common in reptiles than in other groups of animals, many newborns with this condition do not survive to adulthood and it is extremely rare to be born with more than two heads.

SOURCES:

Aldington, Richard and Delano Ames, Transl. (1959.) New Larousse Encyclopedia of Mythology. Hong Kong: Prometheus Press.

Allen, Judy and Jeanne Griffiths. (1979.) The Book of the Dragon. New Jersey: Garnstone Press.

Atsma, Aaron J. (2016, November 5.) Hydra Lernaia. theoi.com. Retrieved from http://www.theoi.com/Ther/DrakonHydra.html.

Bane, Theresa. (2015.) Encyclopedia of Beasts and Monsters in Myth, Legend, and Folklore. Jefferson, North Carolina, USA: McFarland & Company, Inc.

Buxton, Richard, Ed. (2018.) Epic Tales: Greek Myths and Tales. London, UK: Flame Tree Publishing.

Hydra. (2016, March 16.) Encyclopedia Britannica. Retrieved from https://www.britannica.com/topic/Hydra-Greek-mythology.

Jobes, Gertrude. (1962.) Dictionary of Mythology Folklore and Symbols. New York, NY: The Scarecrow Press, Inc.

Rose, Carol. (2000.) Giants, Monsters, and Dragons: An Encyclopedia of Folklore, Legend, and Myth. New York, USA: W. W. Norton & Company, Inc.

Shuker, Karl. (1994.) Dragons: A Natural History. New York: Barnes & Noble Books.

Niles, Doug. (2013.) Dragons: the Myths, Legends, & Lore. Avon, Massachusetts: Adams Media.

Ingersoll, Ernest. (1928, 2014.) Dragons and Dragon Lore. New York: Cosimo Classics.

Cultural Questions:

1. Discuss the similarities and differences of the Christian conversion dragon myths.
2. Buccoleon is an unusual story of a benevolent dragon who travels to western Europe. Discuss Buccoleon's demise and honoring.
3. Cuelebre can only be killed when fed bread with pins or red, hot stones. What are the implications of this?
4. Compare the myth of Cuelebre to Kolowissi.
5. Is Drac a precursor to other dragons? Could the drac myths, with their tales of dragons who can make humans do their bidding, be related to other mythological creatures such as vampires?
6. Discuss the history behind Nant Gwynant. What is the significance of the red and white colors of the dragons?
7. Would you place the Lambton Worm in the Christian conversion myths or as a cautionary tale?
8. Discuss the plethora of healing wells/water in the British Isles.
9. Could the story of the Hydra have influences on Slavic representations of dragons with multiple heads?
10. Discuss the use of dragon stories such as the Tarasque and Graoully in contemporary tourism.
11. Why was Maude added to the Mordiford Wyvern myth in 1864?

BIBLIOGRAPHY

A Chinese Tale. (2014.) Cultural China. Retrieved from traditions.cultural-china.com/en/211T11672T14506.html.

Adelaar, W. F. H. (2006.) Guaraní. Encyclopedia of Language and Linguistics. Oxford: Elsevier.

Adem, Teferi Abate. (2009.) Culture Summary : Tehuelche. New Haven, Conn.: HRAF.

Adrián Ambía, Abel. (1970.) Amaru: Mito i Realidad del Hombre. Lima, Peru: Pukara.

Aftandilian, David. Ed. (2006.) What are the Animals to Us?: Approaches from Science, Religion, Folklore, Literature, and Art. Knoxville, TN: University of Tennessee Press.

Aguirre, Sonia Montecino. (2012.) Chile Precolombino. Museo Chileno De Arte Precolombino.

Akinari, Ueda. (2012.) Ugetsu Monogatari or Tales of Moonlight and Rain (Routledge Revivals): A Complete English Version of the Eighteenth-Century Japanese Collection of Tales of the Supernatural. New York: Routledge.

Aldington, Richard and Delano Ames, Transl. (1959.) New Larousse Encyclopedia of Mythology. Hong Kong: Prometheus Press.

Aldridge, Sally. (1978.) The Peoples of Zambia. London, England: Heineman Educational Books.

Alkha. (2013, December 25.) God Checker. Retrieved from http://www.godchecker.com/pantheon/slavic-mythology.php?deity=ALKHA.

Allen, Judy and Jeanne Griffiths. (1979.) The Book of the Dragon. New Jersey: Garnstone Press.

Allen, Peter J. and Chas. Saunders., Eds. (2013, January 18.) Zmey Gorynich. God Checker. Retrieved from http://www.godchecker.com/pantheon/slavic-mythology.php/deity=ZMEY-GORYNICH.

Álvarez Peña, Alberto Fundación Belenos. (2007.) Elementos de la Antigüedad Celta en la Tradición Oral Asturiana. Etnoarqueologia 13(53), 243.

Amaru. Encyclopaedia Britannica. Retrieved from https://www.britannica.com/topic/Amaru.

Amika, Toshio. (1993.) The Origins of the Grand Shrine of Ise and the Cult of the Sun Goddess Amaterasu Omikami. Japan Review 4, 141-198.

Andre-Driussi, Michael. (1994, 2008.) Lexicon Urthus, 2nd Ed. Albany, CA: Sirius Fiction.

Anjas, A. (1985.) The Baruklinting Dragon. Kawanku Magazine (20). Retrieved from http://www.oocities.org/vienna/5385/nagae.html.

Asbjørnsen, Peter Christen. (1977.) East of the Sun and West of the Moon: Old Tales from the North. Garden City, New York: Doubleday.

Ashley, Mike, ed. (2002.) Arthurian Legends. New Jersey: Castle Books.

Ashliman, D. L. (2010, October 26.) Beowulf: A summary in English prose. Retrieved from www.pitt.edu/~dash/beowulf.html#three.

Ashton, John. (1890, 2014.) Curious Creatures in Zoology. HardPress Publishing.

Aston, William George. (1905.) Shinto: (the Way of the Gods). London, England: Longmans, Green, and Co.

Aston, William George, transl. (1972.) Nihongi: Chronicles of Japan from the Earliest Times to A.D. 697. Book I. Clarendon, VT: Tuttle Publishing Co.

Atsma, Aaron J. (2016, November 5.) Hydra Lernaia. theoi.com. Retrieved from http://www.theoi.com/Ther/DrakonHydra.html.

Azdaha. (2011, August 18.) Encyclopaedia Iranica. Retrieved from http://www.iranicaonline.org/articles/azdaha-dragon- various-kinds#pt1.

Bane, Theresa. (2015.) Encyclopedia of Beasts and Monsters in Myth, Legend, and Folklore. Jefferson, North Carolina, USA: McFarland & Company, Inc.

Barber, Richard and Riches, Anne. (2000.) A Dictionary of Fabulous Beasts. Suffolk, UK: Boydell Press.

Baring-Gould, Sabine. (1863.) Iceland: Its Scenes and Sagas. London: Smith, Elder.

Barton, George A. (1893.) Tiamat. Journal of the American Oriental Society 15, 1-27.

Bathala Myths. Read Legends and Myths. Retrieved from www.read-legends-and-myths.com/bathala-myths.html.

Bathgate, Michael. (2008.) Stranger in the Distance: Pilgrims, Marvels, and the Mapping of the Medieval (Japanese) World. John Hopkins University Press in Medieval Studies. Essays in Medieval Studies 25, 129-144.

Beachum, L. and Horsky, F. (1971.) African Americans. Grade Eight, Unit Three, 8.3 A & B. Comprehensive Social Studies Curriculum for the Inner City. Report. Youngstown Board of Education, OH.Office of Education (DHEW), Washington, DC.

Beckman, Gary. (1982.) The Anatolian Myth of Illuyanka. Journal of the Ancient Near Eastern Society 14, 11-25.

Beckwick, Martha. (1970.) Hawaiian Mythology. Honolulu, Hawaii: University of Hawaii Press.

Beer, Robert. (2003.) The Handbook of Tibetan Buddhist Symbols. Boston, MA, USA: Shambhala, Pub., Inc.

Bellard, A. (1968.) Le Graoully de Metz à la lumière de la paléontologie. Mémoires de l'Académie nationale de Metz, 139-146.

Bellwood, Peter (1987). The Polynesians – Prehistory of an Island People. Thames and Hudson.

Bierlein, J.B. (1999.) Living Myths: How Myth Gives Meaning to Human Experience. New York, NY: Ballantine Books.

Birrell, Anne, tr. (2000.) The Classic of Mountains and Seas. England: Penguin Books.

Biwar. Indonesian Folklore. Retrieved from http://indonesianfolklore.blogspot.com/2008/02/biwar.html.

Blackford, Andy. (2014.) Dragon Tales. Oxford: Oxford University Press.

Blake, Polenth. (2011.) The Dragon Stone-Dragons of Mythology and Fantasy. Retrieved from http://www.polenth.com.

Bolin, Inge. (1998.) Rituals of Respect: The Secret of Survival in the High Peruvian Andes. Austin, TV: University of Texas Press.

Bolla. The Albania Wordpress. Retrieved from https://theealbania.wordpress.com/2014/08/08/albanian-mythology/.

Boney, Jr., Roy. Artist. Retrieved from http://royboney.com/roy-boney-jr-art#/id/i3404041.

Boutet, Michel-Gerald. (2015.) The Great Long Tailed Serpent: An iconographical study of the serpent in Middle Woodland Algonquian culture. Midwestern Epigraphic Society. Retrieved from http://www.midwesternepigraphic.org/ The%20Great%20Long%20Tailed%20Serpent.pdf.

Buchler, Ira R., et al. (2011.) The Rainbow Serpent: A Chromatic Piece. Berlin, Germany: DeGruyter Pub.

Bullion, Alan. (1978.) English Dragon Legends Number 3. Journal of Geomancy 2(3).

Bushell, Raymond. (1977.) Concerning the Walters Collection of Netsuke. The Journal of the Walters Art Gallery 35, 77-85.

Buxton, Richard, Ed. (2018.) Epic Tales: Greek Myths and Tales. London, UK: Flame Tree Publishing.

Byock, Jesse L., trans. (1990.) The Saga of the Volsungs: The Norse Epic of Sigurd the Dragon Slayer. Berkeley and Los Angeles: University of California Press.

Capell, A. (1960.) Language and World View in the Northern Kimberley, Western Australia. Southwestern Journal of Anthropology 16(1), 1-14.

Carnoy, Albert J. (1917.) Iranian Mythology. Boston : Marshall Jones company, 1917.

Carrasco, David. (1982.) Quetzalcoatl and the Irony of Empire: Myths and Prophecies in the Aztec Tradition. Chicago, IL: University of Chicago Press.

Chamberlain, Basil H., transl. (1981.) The Kojiki, Records of Ancient Matters. Clarendon, VT: Tuttle Publishing Co.

Chang, Serena. (2015, June 29.) The legend of the Candle Dragon. The World of Chinese. Retrieved from www. theworldofchinese.com/2015/06/the-legend-of-the-candle-dragon/.

Chavez, Will. (2012, April 17.) Cherokees create artwork using guitars. Cherokee Phoenix. Retrieved from https://www.cherokeephoenix.org/Article/Index/6175.

Cherchenuit, Deleios. (2016, April 1). Dragons around the world. TarValon. Retrieved from https://www.tarvalon.net/ content.php?1744-Dragons-around-the-world.

Chisholm, Hugh, ed. (1911.) Cantabri. Encyclopædia Britannica, 5 (11th ed.). Cambridge University Press.

Chiti, Jorge Fernández. (1997.) Cerámica indígena arqueológica argentina 2, 64.

Chudo Yudo. Mythology Dictionary. Retrieved from www.mythologydictionary.com/baba-yaga-mythology.html.

Coconut Tree. Read Legends and Myths. Retrieved from www.read-legends-and-myths.com/coconut-tree.html.

Colavito, Jason. (2014.) Teshub and the Dragon. Jason and the Argonauts through the Ages. Retrieved from http://www. argonauts-book.com/teshub-and-the-dragon.html.

Colavito, Jason. (2015, December 5.) The Fossil Dragon Bones of Poland's Wawel Cathedral. Jason Colavito. Retrieved from http://www.jasoncolavito.com/blog/the-fossil-dragon-bones-of-polands-wawel-cathedral.

Courlander, Harold. (1996.) A Treasury of African Folklore: The Oral Literature, Traditions, Myths, Legends, Epics, Tales, Recollections, Wisdom, Sayings, and Humor of Africa. New York, USA: Marlowe & Company.

Creedle, William. (2010.) The Otter's Ransom: Moral Accompaniments to Legal Codes in the Icelandic Sagas. Lulu Press, Inc.

Curcó i Pueyo, Jordi. (1996.) "Lo Marraco" i els gegants de Lleida i comarques. Catalonia, Spain: Ajuntament de Lleida.

Curcó i Pueyo, Jordi. (1987.) Els Gegants, Capgrossos i Lo Marraco de Lleida. Collecció La Banqueta: Lleida.

Dalkey, Kara. (2000.) Genpei. New York: Tor Book.

Danver, Steven L. (2013.) Native Peoples of the World: An Encyclopedia of Groups, Cultures and Contemporary Issues. New York, NY: Routledge.

Davis, F. Harland. (1992.) Myths and Legends of Japan. Mineola, N.Y.: Dover Publications.

Davis, Nathan. (1861.) Carthage and Her Remains: Being an Account of the Excavations and Researches on the Site of the Phoenician Metropolis in Africa, and Other Adjacent Places. New York, USA: Harper.

De León, Joanne. (2003.) El dragón y las siete lunas. Japan: Shinseken.

De León, Joanne, and Fure, Masako. (2003.) Otsukisama o nomikonda doragon : Firipin no minwa. Japan: Shinseken.

De Voraigne, Jacobus. (1470, 2012.) The Golden Legend or Lives of the Saints. Princeton, NJ, USA: Princeton University Press.

Dekirk, Ash. (2006.) Dragonlore: From the Archives of the Grey School of Wizardry. New Jersey: Career Press, Inc.

Delehaye, Hippolyte. (1955.) The Legends of the Saints. Dublin: Four Courts Press.

Densmore, S. M. (1976). Mythic Allusion in DH Lawrence's women in Love. Doctoral dissertation. Retrieved from https://macsphere.mcmaster.ca/bitstream/11375/9561/1/fulltext.pdf.

Dixon-Kennedy. (1998.) Encyclopedia of Russian and Slavic Myth and Legend. Santa Barbara, California: ABC-CLIO.

Dobrynya Nikitich and Zmey Gorynych. (2016.) Ural Stone Carving. Retrieved from stonecarving.ru/dobrynya-nikitich-and-zmey-gorynych.html.

Dobson, Kenneth and Arthur Saniotis. (2014.) Dragons: Myth and the Cosmic Powers. Prajna Vihara: Journal of Philosophy and Religion 15(1), 85-99.

Douglas, Lachlan. (2013, October 18.) Chinese Myth of the Candle Dragon. Ancient Chinese Myths and Legends. Retrieved from www.ancientchina7b.weebly.com/1/post/2013/10/chinese-myth-of-the-candle-dragon.html.

Dragons & Serpents in Sussex. Sussexarch.org.uk. Retrieved from http://www.sussexarch.org.uk/saaf/dragon.html.

Dragon of Ghent. Sacred Texts. Retrieved from www.sacredtexts.com/etc/tl/tl08.htm.

Dragons in Mythology, Legend & History, Part 2. (2014, July 6.) Myth Beliefs. Retrieved from http://www.mythbeliefs. info/2014/07/dragons-in-mythology-legend-history.html.

Dreamtime Stories - the Rainbow Serpent. (2001, September 26.) Australia Lesson Activities. Retrieved from http://www. expedition360.com/australia_lessons_literacy/2001/09/dreamtime_stories_the_rainbow.html.

Druett, Joan. (2000.) She-Captains: Heroines and Hellions of the Sea. New York: Simon & Schuster.

Dunton, Larkin. (1896). The World and Its People. Silver, Burdett.

Edmonds, Margot and Ella Clark. (1989.) Voices of the Winds: Native American Legends. Castle Books.

Edrick, Vann. Amaruca and America. Controversial Files. Retrieved from controversialfiles.blogspot.com/2013/01/amaruca-and-america.html.

Egyptian Culture. (2013, August 1.) Ancient History Encyclopedia. Retrieved from http://www.ancient.eu/Egyptian_Culture/.

Elkin, Z. P. (1930.) The Rainbow-Serpent Myth in North-West Australia. Oceania 1(3), 349-352.

Elsie, Robert. (2001.) A Dictionary of Albanian Religion, Mythology and Folk Culture. Washington Square, New York: New York University Press.

Emerson, Nathaniel B. Pele and Hiiaka. Retrieved from https://archive.org/stream/pelehiiakamythfr00emeriala/ pelehiiakamythfr00emeriala_djvu.txt.

Emperaire, J. and Laming, A. (1954.) La Grotte Du Mylodon (Patagonie Occidentale). Journal de la Société des américanistes, NOUVELLE SÉRIE 43, 173- 205.

Erdoes, Richard and Alfonso Ortiz. (1984.) American Indian Myths and Legends. New York, NY, USA: Pantheon Books.

Evans, Jeff. (2009.) Nga Waka O Nehera: The First Voyaging Canoes. New Zealand: Libro International.

Evans, Jonathan. (2008.) Dragons. London, UK: Apple Press.

Exploring Legendary German History on Dragon Rock. (2007, August 2.) Retrieved from http://www.dw.com/en/exploring-legendary-german-history-on-dragon-rock/a-2715873.

Fegan, Brian. (1983.) Some Notes on Alfred McCoy, 'Baylan: Animist Religion and Philippine Peasant Ideology.' Philippine Quarterly of Culture and Society 11(2/3), 212-216.

Floyd, Randall. (1998, September 20.) Disappearances feed Grootslang legend. Augusta Chronicle, Retrieved from http:// old.chronicle.augusta.com/stories/1998/09/20/ent_239491.shtml.

Foster, Michael Dylan. (1998.) The Metamorphosis of the Kappa: Transformation of Folklore to Folklorism in Japan. Asian Folklore Studies 57, 1-24.

Fox, C.E. and Drew, F. H. (1915.) Beliefs and Tales of San Cristoval (Solomon Islands). The Journal of the Royal Anthropological Institute of Great Britain and Ireland 45, 131-185.

Fredericks, Deby. (2012, March 22.) The Four Dragons, a Chinese folk story. Wyrmflight. Retrieved from https://wyrmflight.wordpress.com/2012/03/22/the-four-dragons-a-chinese-folk-story/.

Freeman, Richard. (2006.) The Drac. Myths. Retrieved from http://myths.e2bn.org/mythsandlegends/userstory8588-the-drac.html#top.

Freeman, Richard. Mordiford Wyvern. Mysteriousbritain.co.uk. Retrieved from http://www.mysteriousbritain.co.uk/england/hereford-and-worcestershire/legends/mordiford-wyvern.html.

Friedman, Amy and Meredith Johnson. (2001, January 28.) The Dragon's Pearl (An Ancient Chinese Legend). UExpress: Tell Me a Story. Retrieved from http://www.uexpress.com/tell-me-a-story/2001/1/28/the-dragons-pearl- an-ancient-chinese.

Georgieva, Ivanička (1985). Bulgarian Mythology. Translated by Vessela Zhelyazkova. Svyat Publishers.

Gibson, Chris. (1997.) 'Nitmiluk': Song-sites and Strategies for Aboriginal Empowerment. Land and Identity: Proceedings of the Nineteenth Annual Conference; Journal of the Association for the Study of Australian Literature, University of New England, Armidale, September 27–30, Sydney: University of New England Press: 161-167.

Giordano, John T. (2005.) Kinnari: On the Space between Traditional and Corporate Myths. Prajna Vihara 6(1), 90-107.

Gong Gong (the God of Water). (2014.) Cultural China. Retrieved from http://traditions.cultural-china.com/en/13Traditions1394.html.

Gong Gong, Nuwa, and the Fragile Nature of Life. (2011, May 19.) ferrebeekeeper. Retrieved from https://ferrebeekeeper.wordpress.com/2011/05/19/gong-gong-nuwa-and-the-fragile-nature-of-life/.

Goodsoul, Sofia. (2015.) Nian, the lunar dragon: a famous Chinese legend retold by Sofia Goodsoul and Marina Kite. Clayton South, Vic. Volya Press.

Gordeev, Nikolai P. (2017.) Snakes in the Ritual Systems of Various Peoples. Anthropology & archeology of Eurasia 56(1-2), 93-121.

Gould, Charles. (1886.) Mythical Monsters. New York: Cosimo Classics.

Grafman, R. (1972.) Bringing Tiamat to Earth. Israel Exploration Journal 22(1), 47-49.

Grantham, Bill. (2002.) Creation Myths and Legends of the Creek Indians. University Press of Florida.

Grayson, James H. (2013.) Korea - A Religious History. New York: Routledge.

Green, Roger Lancelyn. (1973.) A Book of Dragons. New York: Penguin Books.

Griffs, William Elliott. (1919.) Belgian Fairy Tales. Thomas Y. Crowell Co.

Gulden Drak. Brouwerij Van Steenberge. Retrieved from www.vansteenberge.com/en/our-beer/gulden-draak/gulden-draak.

Hall, Lesslie, trans. (1892.) Beowulf: An Anglo-Saxon Epic Poem, Translated From The Heyne-Socin. Boston, New York, Chicago: D. C. Heath & Co., Publishers.

Haney, Enoch Kelly. Legend of the Snake Clan. Seminole Nation Museum, Wewoka OK. Retrieved from https://www.seminolenationmuseum.org/m.blog/23/legend-of-the-snake-clan.

Hargreaves, Joyce. (2009.) A Little History of Dragons. New York: Walker & Company.

Harner, Michael J. (1973.) The Jivaro: People of the Sacred Waterfalls. Garden City, NY: Anchor Books.

Harner, Michael J. (1974.) The Sound of Rushing Water. In Native South Americans. Patricia J. Lyon, Ed. Boston: Little, Brown and Company.

Harple, T.S. (2001). Controlling the Dragon: An ethno-historical analysis of social engagement among the Kamoro of South-West New Guinea (Indonesian Papua/ Irian Jaya). Unpublished PhD thesis, Australian National University, Canberra.

Harrington, Jo. (2014, October 1.) Krakow and the Legend of the Wawel Dragon. Retrieved from www.wizzley.com/krakow-dragon.

Hawai'i's Mo'o: Tiny Gecko, Seductive Woman, or Water Dragon? CryptoVille. Retrieved from http://visitcryptoville.com/2015/09/30/hawaiis-moo-tiny-gecko-seductive-woman-or-water-dragon/.

Hayter, Maud Goodenough. (1957, 2000.) Folklore and Fairy Tales of the Canterbury Maoris. Christchurch, New Zealand: Cadsonbury Publications.

Heuvelmans, Bernard. (1958, 1995.) On the Track of Unknown Animals. New York: Routledge.

Hidden Inca Tours: Explore Mankind's Hidden History. (2017) Retrieved from http://hiddenincatours.com.

Hirsch, Emil, Kaufmann Kohler, Solomon Schechter, Isaac Broyde. Leviathan and Behemoth. Jewish Encyclopedia. Retrieved from http://www.jewishencyclopedia.com/articles/9841-leviathan-and-behemoth.

HN. (2014, November 24.) Soutien d'un Français d'origine africaine. Courrier des lecteurs. Retrieved from https://translate.google.com/translate?hl=en&sl=fr&tl=en&u=http%3A%2F%2Fwww. egaliteetreconciliation.fr%2FCourrier-des-lecteurs-29304.html.

Hodges, Margaret. (1984.) Saint George and the Dragon: A Golden Legend. Boston, MA: Little, Brown, and Company.

Holloway, April. (2015, April 22.) The Gods of Creation and Legendary Beasts of the Guarani. Ancient Origins: Reconstructing the Story of Humanity's Past. Retrieved from http://www.ancient-origins.net/myths-legends- americas/gods-creation-and-legendary-beasts-guarani-002937.

Holloway, April. (2014, January 31.) The Origin of Lunar New Year and the Legend of Nian. Ancient Origins: Reconstructing the Story of Humanity's Past. Retrieved from ancient-origins.net/myths-legends-asia/origin-lunar-new- year-and-legend-nian-001289.

Hubbs, Joanna. (1993.) Mother Russia: The Feminine Myth in Russian Culture. Bloomington and Indianapolis: Indiana University Press.

Hudson, Angela Pulley. (2010.) Creek Paths and Federal Roads: Indians, Settlers, and Slaves and the Making of the American South. North Carolina: University of North Carolina Press.

Hydra. (2016, March 16.) Encyclopedia Britannica. Retrieved from https://www.britannica.com/topic/Hydra-Greek-mythology.

Inca Trail Treks. Kondor Path Tours. Retrieved from www.kondorpathtours.com.

Ingersoll, Ernest. (1928, 2014.) Dragons and Dragon Lore. New York: Cosimo Classics.

Ivens, Walter. (1934.) The Diversity of Culture in Melanesia. The Journal of the Royal Anthropological Institute of Great Britain and Ireland 64, 45-56

Iwanci. The Strange Myth World. Retrieved from https://thestrangemythworld.com/2015/05/20/iwanci/.

Iya, Palmo R. (2016.) EL RENASCIMENTO: UNVEILING THE METAPHORICAL MEANING OF BATHALA. Philippine Association for the Sociology of Religion Journal 2, 1, 1-24.

Jacobs, Joseph. (1894.) More English Folktales. London: David Nutt and Co.

Jacobsen, Thorkild. (1977.) Mesopotamia. In Before Philosophy: The Intellectual Adventure of Ancient Man. Chicago, IL: University of Chicago Press.

Jacobsen, Thorkild. (1968.) The Battle between Marduk and Tiamat. Journal of the American Oriental Society 88(1), 104-108.

Jauregui, Andres. (2014, September 24.) Lagarfjótsormur, Iceland's Legendary Lake Monster, Caught On Tape, Panel Says. Huffngton Post. Retrieved from https://www.huffngtonpost.com/2014/09/24/lagarfjotsormur-sea-monster-video_n_5875970.html.

JHTI. (2002.) Nihon Shoki," Japanese Historical Text Initiative (JHTI). Berkeley, CA: University of Berkeley, English tr. Aston.

Jobes, Gertrude. (1962.) Dictionary of Mythology Folklore and Symbols. New York, NY: The Scarecrow Press, Inc.

Jocano, F. Landa. (1969.) Outline of Philippine Mythology. Manila: Centro Escolar University Research and Development Center.

Johns, Andreas. (2004.) Baba Yaga: the Ambiguous Mother and Witch of the Russian Folktale. Peter Lang.

Joly, Henri L. (1908.) Legend in Japanese Art: A Description of Historical Episodes, Legendary Characters, Folk- lore Myths, Religious Symbolism. London: John Lane, The Bodley Head.

Jones, Gwyn and Thomas Jones. (1949, 1991.) The Mabinogion. London: Everyman.

Jormungand. (2016.) Norse Mythology. Retrieved from http://norse-mythology.org/gods-and-creatures/giants/jormungand/.

Juhl, R. A., E. G. Mardon, and A. A. Mardon. (2007.) Documentary Evidence for the Apparition of a Comet in Late 593 and Early 594 AD. Lunar and Planetary Science XXXVIII, 1184.

Kaberry, Phyllis M. (1936.) Spirit-Children and Spirit-Centres of the North Kimberley Division, West Australia. Oceania 6(4), 392-400.

Kaignavongsa, Xay and Hugh Fincher. (1993.) Legends of the Lao: A Compilation of Legends and other Folklore of the Lao People. USA: Geodata Systems, pp. 23-26.

Kaleoikapolialoha. (1998.) Tales from the Enchanted Isles Hawaii. Honolulu, Hawaii: Ka ‘imi Pono Press.

Kamminga, Johan. (1999.) Prehistory of Australia. Australia: Allen and Unwin Pub.

Kawbawgam, Charles, Kawbawgam, Charlotte, LePique, Jacques, Kidder, Homer H., and Bourgeois, Arthur P. (1994.) Ojibwa narratives of Charles and Charlotte Kawbawgam and Jacques LePique, 1893-1895. Detroit : Wayne State University Press.

Ke, Yuan. (1991.) Dragons and Dynasties: An Introduction to Chinese Mythology. Translated by Kim Echim and Nie Zhixiong, London: Penguin.

Keally, Charles T. (2009, October 13.) Japanese Paleolithic Period. Retrieved from http://www.t-net.ne.jp/~keally/palaeol. html/

Keane, Basil. ‘Taniwha’, Te Ara. The Encyclopedia of New Zealand. Retrieved from http://www.TeAra.govt.nz/en/taniwha.

Kendall, Timothy. (2015, March 29). III. G. Jebel Barkal in the Book of the Dead. Black Drago. Retrieved from http://www. blackdrago.com/fame/mehen.htm.

King, Leonard. (1902, 2014.) The Seven Tablets of Creation. CreateSpace Publishing Platform.

King, William R. (1987.) Dionysos Among the Mesas: The Water Serpent Puppet Play of the Hopi Indians. American Indian Culture and Research Journal 11(3), 17-49.

Krakow info-Dragon Den. krakow-info.com. Retrieved from http://www.krakow-info.com/smocza.htm.

Kramer, Samuel Noah. (1944, 2007.) Sumerian Mythology: A Study of Spiritual and Literary Achievement in the Third Millennium B.C. London: Forgotten Books.

Krishnamurphy, K. (1984.) Mythical Animals in Indian Art. New Delhi, India: Abhinv Publications.

Krupa, Viktor. (2002.) A Mythological Metamorphosis: Snake or Eel? Asian and African Studies 11(1), 9-14.

Lane, Bertha Palmer. (2009.) Tower Legends. Abela Pub. Ltd.

Lang, Andrew. (1897.) The Pink Fairy Book. Dover: Dover Publications.

Laparenok, Leonid. Prominent Russians: Vladimir I. Russiapedia (Get to know Russia better). Retrieved from https://russiapedia.rt.com/prominent-russians/history-and-mythology/vladimir-i/.

Latini, David, Ed. (2018.) Epic Tales: Chinese Myths and Tales. London, UK: Flame Tree Publishing.

Lawson, Julie. The Dragon's Pearl. Unit 1 No Turning Back. Retrieved from http://ditter91.weebly.com/uploads/5/1/3/7/51371393/the_dragons_pearl.pdf.

Le, C. N. Vietnam: Early History and Legend. Asian-Nation. The Landscape of Asian America. Retrieved from http://www. asian-nation.org/vietnam-history.shtml.

Leandro M. Pérez, Néstor Toledo, Sergio F. Vizcaíno, M. Susana Bargo. (2018). Los restos tegumentariosde perezosos terrestres (Xenarthra, Folivora) de Última Esperanza (Chile). Cronología de los reportes, origen y ubicación actual. Publicación Electrónica de la Asociación Paleontológica Argentina 18(1), 1–21.

Legend of the Graoully Dragon and Saint Clement of Metz. (2012, December 12.) Green Mountain College: The Rope Swing. Retrieved from https://theropeswingblog.wordpress.com/2012/12/29/legend-of-the-graoully-dragon-and-

saint-clement-of-metz/.

Legends. Visit Me TZ: Office de Tourisme Metz Métropole. Retrieved from http://www.tourisme-metz.com/en/legends.

Legg, Gerald. (2006.) Dragons. Great Britain: Book House.

Liu, Fenggui and Feng, Zhaodong. (2012.) A dramatic climatic transition at ~4000 cal. yr BP and its cultural responses in Chinese cultural domains. The Holocene 22(10), 1181–1197.

Loh-Hagan, Virginia. (expected 2019.) Nian, the Chinese New Year Dragon. Ann Arbor, Michigan: Sleeping Bear Press.

Longwitton (Northumberland). Retrieved from http:www.wyrm.org.uk/ukdracs/longwitton.html.

Ludvik, Catherine. (2001.) From Sarasvati to Benzaiten. Ph.D. Thesis, University of Toronto, National Library of Canada.

Lurker, Manfred. (2015.) A Dictionary of Gods and Goddesses, Devils and Demons. London and New York: Routledge.

Lynch, Patricia Ann. (2004.) African Mythology A to Z. New York, USA: Chelsea House Publishers.

Mannering, May. (1866). Drachenfels; Or, The Dragon's Rock. Student & Schoolmate: An Illustrated Monthly for all our Boys & Girls 17(6), 218-221.

Manu, Moke. (2006.) Hawaiian Fishing Traditions. Honolulu, Hawaii: Kalamaku Press.

Manora: A Thai Legend. Manora Property. Retrieved from http://www.manoraproperty.com/136-manora-a-thai-legend.

Maori Culture. 100% Pure New Zealand. Retrieved from http://www.newzealand.com/my/maori-culture.

Mapuche. Minority Rights. Retrieved from https://minorityrights.org/minorities/mapuche-2/.

Mark, Joshua J. (2016, January 9.) The Hittites. Retrieved from www.ancient.eu/hittite.

McCall, Gerrie. (2007.) Dragons: Fearsome Monsters from Myth and Fiction. New York, NY: Tangerine Press.

McCall, Gerrie and Regan, Lisa. (2011.) Monsters and Myths: Dragons and Serpents. New York, NY: Gareth Stevens Publishing.

McCall, Henrietta. (1990.) Mesopotamian Myths. London, UK: British Museum Press.

McCormick, Kylie. (2017, October 9.) Aganua. Dragons of Fame. Retrieved from http://www.blackdrago.com/fame/agunua.htm.

McCormick, Kylie. (2017, October 9). Ananta. Dragons of Fame. Retrieved from http://www.blackdrago.com/ fame/ananta.htm.

McCormick, Kylie. (2003, September 9.) Chudo Yudo. Dragons of Fame. Retrieved from ww.blackdrago.com/fame/chudoyudo.htm.

McCormick, Kylie. (2012, November 18.) Dragon History. Dragons of Fame. Retrieved from www.blackdrago.com/history/outline.htm.

McCormick, Kylie. (2012, November 18.) Dragon Pearl. Black Drago. Retrieved from www.blackdrago.com/history/eastpearl.htm.

McCormick, Kylie. (2017, October 9.) Gandarewa. Dragons of Fame. Retrieved from http://www.blackdrago.com/fame/ gandareva.htm.

McCormick, Kylie. (2013, September 9.) Iemisch. Dragons of Fame. Retrieved from www.blackdrago.com/fame/iemisch.htm.

McCormick, Kylie. (2013, September 9.) Ihuaivulu. Dragons of Fame. Retrieved from www.blackdrago.com/fame/ihuaivulu.

McCormick, Kylie. (2013, September 9.) Iwanci. Dragons of Fame. Retrieved from www.blackdrago.com/fame/Iwanci.

McCormick, Kylie. (2013, September 9.) Nidhogg. Dragons of Fame. Retrieved from www.blackdrago.com/fame/nidhogg.htm.

McCormick, Kylie. (2013, September 9.) Walutahanga. Dragons of Fame. Retrieved from www.blackdrago.com/fame/ walutahanga.htm.

McCullogh, Joseph. (2013.) From Beowulf to St. George. Bloomsbury Publications.

McLeish, Kenneth. (1996). Myth: Myths and Legends of the World Explored. New York, NY: Faces On File, Inc.

Megatherium. Prehistoric Wildlife. Retrieved from http://www.prehistoric-wildlife.com/species/m/megatherium.html.

Messer, Ron. (1989.) A Structuralist's View of an Indian Creation Myth. Anthropologica 31(2), 195-235.

Methvin, Rev. J.J. (1927, December.) Legend of the Tie-Snakes. Chronicles of Oklahoma 5(4).

Métraux, Alfred. (1948, 1998). The Guarani. New Haven, Conn.: HRAF.

Meurger, Michel and Claude Gagnon. (1998.) Lake Monster Traditions: A Cross-Cultural Analysis. London: Fortean Times.

Mills, Barbara J. and T.J.Ferguson. (2008.) Animate Objects: Shell Trumpets and Ritual Networks in the Greater Southwest. Journal of Archaeological Method and Theory 15, 342.

Mitchell, Deborah. King of the Waters: The Legend of the Horned Water Serpent. Southeastern Oklahoma State University. Retrieved from http://www.se.edu/nas/files/2013/03/2ndsymposiumpart5.pdf.

Mo'o: Dragons of Hawai'i. (2013, January 10.) Uncanny Hawaii: The Unconventional Guide to Hawaii. Retrieved from http:// uncannyhawaii.com/mo-o-dragons-of-hawaii-vampire-drake/.

Moffat, Charles Alexander. Gaasyendietha. Sea Serpents of Canada. Retrieved from http://www.lilith-ezine.com/ articles/2005/canadian_seaserpents.html.

Monte, Richard. (2008.) The Dragon of Krakow. London, UK: Frances Lincoln.

Mooney, James. (1900.) History, Myths, and Sacred Formulas of the Cherokees. 19th Annual Report of Bureau of American Ethnology 1897-98, Part I.

Morris, Desmond and Ramona. (1965.) Men and Snakes. New York: McGraw Hill Book Co.

Mosby, Kristina and Jake Hicks. Mount Pilatus. Lucerne Switzerland. Retrieved from https://lucerneswitzerland.weebly.com/mount-pilatus.html.

Mythic Hawaii. Ancient Hawaiian History Origins of the Hawaiian Islands, Culture and Natives. Retrieved from http:// mythichawaii.com/hawaiian-history-culture.htm.

Nanibush, Wanda. (2018.) Nanabozho's Sisters. Canadian Art 35(3), 37-37.

Napier, Susan J. (2005.) Anime from Akira to Howl's Moving Castle: Experiencing Japanese Animation. Palgrave Macmillan.

Narciso Rosicrán Colman. (1929.) Ñande Ypy Kuéra ("Nuestros Antepasados"). BibliotecaVirtual del Paraguay. Retrieved from https://web.archive.org/web/20070930224211/http://www.bvp.org.py/biblio_htm/colman/indice.htm.

Native Languages of the Americas: Guarani Indian Legends, Stories, and Myths. (2015.) Native Languages. Retrieved from http://www.native-languages.org/guarani-legends.htm.

Nelson, E. W. (1900) The Eskimo about Bering Strait. Extract from the Eighteenth Annual Report of the Bureau of American Ethnology. Government Printing Office, Washington.

Nery, Peter Solis. (2012.) Love in the Time of the Bakunawa. United States: CreateSpace Independent Publishing Platform.

Nevins, Scott. (2015, April 26.) The 'Dragon-Slayer' Saints of the Eastern Orthodox Church. Scott Nevins Memorial. Retrieved from https://scottnevinssuicide.wordpress.com/2015/04/26/the-dragon-slayer-saints-of-

the-eastern-orthodox-church/.

News24. (2015, October 28.) Angry river god blamed for parched Kariba. Retrieved from https://www.news24.com/Africa/

Zimbabwe/Angry-river-god-blamed-for-parched-Kariba-20151028.

Nguyen, Dieu Thi. (2013.) A mythographical journey to modernity: The textual and symbolic transformations of the Hùng Kings founding myths. Journal of Southeast Asian Studies 44(2), 315-337.

Nickel, Helmut. (1991.) The Dragon and the Pearl. Metropolitan Museum Journal 26, 139-146.

Nidhogg. Draconika. Retrieved from http://www.draconika.com/legends/nidhogg.php.

Nidhogg. (2012, January 26.) tobiasmastgrave.wordpress. Retrieved from https://tobiasmastgrave.wordpress.com/tag/nidhogg/.

Niles, Doug. (2013.) Dragons: the Myths, Legends, & Lore. Avon, Massachusetts: Adams Media.

Norwich Dragon Festival. (2014.) Norwich Dragon Festival Education Pack. World Art Collections Exhibitions, Sainsbury Centre for Visual Arts. Retrieved from http://www.heritagecity.org/user_files/downloads/education-pack-vf-web.pdf.

Nyaminyami, River God and Spirit of the Zambezi River, Zimbabwe. Zambezi Safari & Travel Company. Retrieved from

https://www.zambezi.com/blog/2013/nyaminyami-spirit-zambezi-river/.

Nyaminyami – the Kariba Legend. Zambezi Safari & Travel Company. Retrieved from https://www.zambezi.com/ blog/2011/nyaminyami-the-kariba-legend/.

Nyamukondiwa, Walter. (2018, March 2.) ZTA to Launch Nyaminyami Festival. All Africa, Retrieved from https://allafrica. com/stories/201803020581.html.

Ō, no Yasumaro. (2014.) The Kojiki : an account of ancient matters. Translated by Gustav Heldt, New York: Columbia University Press.

O'Callaghan, Regan. (2011, June 9.) Horomatangi. Regan O'Callaghan. Retrieved from http://www.reganocallaghan. com/?p=123ori/Maori-Myths-Legend.

O'Leary, Michael. (2013). Sussex Folk Tales. The History Press.

O'Neill, Patt. (2007, June 13.) Glossary of Terminology of the Shamanic & Ceremonial Traditions of the Inca Medicine Lineage. Inca Glossary. Retrieved from http://www.incaglossary.org/a.html.

Of cooks, pirates, and dragons: Saint Martha. (2011, July 29.) A Nun's Life Ministry. Retrieved from https://anunslife.org/blog/nun-talk/of-cooks-pirates-and-dragons-saint-martha.

Origins of Maud and the Dragon. (2006.) Myths and Legends. Retrieved from http://myths.e2bn.org/mythsandlegends/origins1353-maud-and-the-dragon.html#top.

Ouwehand, Cornelius. (1958-1959.) Some Notes on the God Susa-no. Monumenta Nipponica 14(3⁄4), 384-407.

Page, Willie F., and R. Hunt Davis, eds. (2005.) "Ougadou-Bida." In Encyclopedia of African History and Culture: African Kingdoms (500 to 1500), vol. 2. New York, USA: Facts on File, Inc.

Pai Lung. The Sovereign Lair. Retrieved from www.angelfire.com/dragon2/thoth/pailung.html.

Pai Lung. (2015.) Mythology Dictionary. Retrieved from www.mythologydictionary.com/pai-lung-mythology.html.

Paraguayan Myths. (2006.) Project Paraguay. Retrieved from http://www.projectparaguay.com/myths.htm.

Parrinder, Geoffrey. (1961.) West African Religion: A Study of the Beliefs and Practices of Akan, Ewe, Yoruba, Ibo, and Kindred Peoples. London: Epworth Press.

Parsons, Elsie Clews. (1923.) The Origin Myth of Zuni. The Journal of American Folklore 36(140), 135-162.

Penncavage, Michael. (2010.) The Bakunawa. In Dragon's Lure: Legends of a New Age. Danielle Ackley-McPhail, Jennifer Ross, and Jeffrey Lyman, Eds. Howell, New Jersey: Dark Quest, LLC.

Pereira, Vicente Cretton. (2016.) Our Father, Our Owner: Master Relations between the Mbya Guaraní. Mana 22(3), 737-764.

Peterson, Joseph H. (1995.) Avesta: Khorda Avesta. Avesta.org. Retrieved from http://www.avesta.org/ka/yt5sbe.htm.

Piccardi, Luigi and W.Bruce Masse. (2007.) Myth and Geology. London: Geological Society of London.

Pike, A.W.G. & Hoffmann, D.L. & García-Diez, Marcos & Pettitt, Paul & González, José Javier & De Balbin-Behrmann, Rodrigo & González-Sainz, C & de las Heras, C & Lasheras, J.A. & Montes, R & Zilhão, João. (2012). U-Series Dating of Paleolithic Art in 11 Caves in Spain. Science 336(6087), 1409-1413.

Plowright, Poh Sim. (1998.) The Art of Manora: an Ancient Tale of Feminine Power Preserved in South-East Asian Theater. New Theatre Quarterly 14(56), 373-394.

Poignant, Roslyn. (1967.) Oceanic Mythology: The Myths of Polynesia, Micronesia, Melanesia, Australia. London: Paul Hamlyn.

Polish Legends, Myths and Stories. Anglik.net. Retrieved from http://www.anglik.net/polish_legends_dragon.htm.

Portal, Claire. (2013.) When Unremarkable Landscapes Receive Heritage Status: Considerations Based on Case Studies in the Pays de la Loire (France). L'Espace géographique 42, 213-226.

Pra Suthon-Manora. Translated by Marion Davies. Gotoknow. Retrieved from https://www.gotoknow.org/posts/458723.

Prehistory and Ancient History. National Museum of Korea. Retrieved from

https://www.museum.go.kr/site/eng/showroom/list/760?showroomCode=DM0002.

Prifti, Peter R. and Biberaj, Elez. (2019, February 22.) Encyclopaedia Britannica. Retrieved from https://www.britannica.com/place/Albania.

Princess Manora. Angelgenie. Retrieved from www.angelgenie.com/index.php?cPath=20.21.

Privat, Jean-Marie, dir. (2006.) Dragons entre sciences et fiction. Paris, CNRS Éd.

Qingdao China Guide. (2016.) Retrieved from qingdaochinaguide.com.

Quetzalcoatl. (2013, August 1.) Ancient History Encyclopedia. Retrieved from http://www.ancient.eu/Quetzalcoatl/.

Radcliffe-Brown, A. R. (1926.) The Rainbow-Serpent Myth of Australia. The Journal of the Royal Anthropological Institute of Great Britain and Ireland 56, 19-25.

Radcliffe-Brown, A. B. (1930.) The Rainbow-Serpent Myth in South-East Australia. Oceania 1(3), 342-347.

Radhakrishnan, Reeja. (2014, June 27.) Tchang and the Dragon's Pearl. The New Indian Express. Retrieved from www.newindianexpress.com.

Rafferty, Patrick. (2007, March 9.) Visayan-English dictionary. The United States and its Territories: 1870- 1925: The Age of Imperialism. Retrieved from http://quod.lib.umich.edu/p/philamer/ACK6070.0001.001/18?rgn=full+text;view=image;q1=bakunawa.

Ralston, William Ralston Shedden. (1880.) Russian Folk-tales. R. Worthington.

Reed, A. W. (1963.) Treasury of Maori Folklore. Hong Kong: Dai Nippon Printing Co.

Reed, A. W. (1964.) Maori Fables and Legendary Tales. Great Britain: C. Tinling & Co. Ltd.

Reed, A. W., and Calman, Ross. (2008). Taniwha, Giants, and Supernatural Creatures: He Taniwha, He Tipua, He Patupaiarehe. North Shore, New Zealand: Raupo.

Resture, Jane. (2009, October 7.) Melanesian Mythology: Solomon Islands. Jane's Oceania Home Page. Retrieved from http://www.janeresture.com/melanesia_myths/solomons.htm.

Revelation 12:3. Bible.hub. Retrieved from http://biblehub.com/revelation/12-3.htm.

Richtersveld Route. (2015.) Tourism Route North South. Retrieved from http://www.south-north. co.za/rich_rt.html.

Roblee, Mark. (2018.) Performing Circles in Ancient Egypt From Mehen to Ouroboros. Preternature: Critical and Historical Studies on the Preternatural, 7(2), 133-153.

Rose, Carol. (2000.) Giants, Monsters, and Dragons: An Encyclopedia of Folklore, Legend, and Myth. New York, USA: W. W. Norton & Company, Inc.

Ruland, Wilhelm. The Drachenfels. Retrieved from http://www.kellscraft.com/LegendsRhine/legendsrhine081.html.

Rustan, Mario. Biwar Kills a Dragon. Indonesian Myth. Retrieved from http://www.st.rim.or.jp/~cycle/MYdragonE.HTML.

Sá, Lúcia. (2004) Rain Forest Literatures: Amazonian Texts and Latin American Culture. Minneapolis, Minnesota: University of Minnesota Press.

San Buenaventura, Mariejoy. (2018.) Book Review: Mythological Woman and the Prose Poem in Barbara Jane Reyes's Diwata. Ramkhambaeng University Journal Humanities Edition 37, 1, 193-210.

Sanday, Peggy Reeves. (2007.) Aboriginal Paintings of the Wolfe Creek Crater: Track of the Rainbow Serpent. Philadelphia, Pennsylvania: University of Pennsylvania Museum of Archaeology and Anthropology.

Sanders, Tao Tao Liu. (1983.) Dragons, Gods & Spirits from Chinese Mythology. New York: Schocken Books.

Schorn, Brittany. (2018.) Epic Tales: Norse Myths and Tales. London, UK: Flame Tree Publishing.

Schultz, Jack Maurice. (2008.) The Seminole Baptist Churches of Oklahoma: Maintaining a Traditional Community. Norman, OK: University of Oklahoma Press.

Sea Lamprey. British Sea Fishing. Retrieved from http://britishseafshing.co.uk/sea-lamprey/.

Sharrock, Peter D. (2015.) Serpent-enthroned Buddha of Angkor. Marg, A Magazine of the Arts, 20.

Sheng, Jim. (2012, June 19.) The Foolish Dragon. Chinese Aesop: Fairy Tales, Folk Tales, Fables, Myths, Legends, and Historical Stories. Retrieved from chineseaesop.blogspot.com/2012/06/foolish-dragon.html.

Shuker, Karl. (1994.) Dragons: A Natural History. New York: Barnes & Noble Books.

Shung Ye Museum of Formosan Aborigines Guidebook. (2017.)

Sihombing, Ronny Sahputra, Palupi, Victoria Usadya, and Ragawanti, Debora Tri. (2013). Identification Of Six Elements Of Narrative Used By Third Grade Of Elementary School Students Of Bethany School In Rewriting The Story Of Rawa Pening. Satya Wacana Christian University Institutional Repository. Retrieved from http:// repository.uksw.edu/handle/123456789/3402.

Simpson, Jacqueline. (1972). The Water-Snake of Lagarfjot. Icelandic Folktales and Legends. Berkeley and Los Angeles: University of California Press.

Simpson, Jacqueline. Tarasque. Retrieved from folklore-society.com/miscellany/tarasque.

Slaveykov, Racho. (2014.) Bulgarian Folk Traditions and Beliefs. Sofja: Asenevci.

Smith, Michael E. (2003.) The Aztecs 2nd Ed. Malden, MA: Blackwell Pub. Ltd.

Smith, Richard Gordon. (1918.) Ancient Tales and Folk-lore of Japan. Montana: Kessinger Publishers.

Smith, S. (2011). Generative landscapes: the step mountain motif in Tiwanaku iconography. Ancient America, 12, 1–69.

Somerville, Angus A. and R. Andrew McDonald, Eds. (2010.) The Viking Age: A Reader. Toronto: University of Toronto Press.

Sonofskankware. (2012, August 20.) Drac. Mythical Creatures Guide. Retrieved from http://www.mythicalcreaturesguide.com/page/Drac.

Steele, Paul Richard and Catherine J. Allen. (2004.) Handbook of Inca Mythology. Santa Barbara, CA: ABC-CLIO.

Stookey, Lorena Laura. (2003.) Thematic Guide to World Mythology. Santa Barbara, CA: Greenwood Publishing Group.

Stop digging Down Under? A zinc mine in Australia meets resistance among Aborigines concerned about the environment and a 'rainbow serpent.'. (2006). Christian Science Monitor 16, 4.

Stothers, Richard B. (2004.) Ancient Scientific Basis of the 'Great Serpent' from Historical Evidence. Isis 95(2), 220-238.

Swann, Brian, Ed. (2004.) Voices from Four Directions: Contemporary Translations of the Native Literatures of North America (Native Literatures of the Americas). Lincoln: University of Nebraska Press.

Tabada, Mayette Q. (2003, March 2.) Tabada: Why doctors become nurses. Babaylan Files. Retrieved from http://babaylanfiles.blogspot.com/2009/07/conversations-tabada-why-doctors-become.html.

Tacon, Paul, Meredith Wilson and Christopher Chippindale. (1996) Birth of the Rainbow Serpent in Arnhem land rock art and oral history. Archaeology in Oceania 31(3), 103-124.

Taiwan volcano: active, dormant, or what? (2009, November.) The Volcanism Blog. Retrieved from

https://volcanism.wordpress.com/2009/11/02/taiwan-volcano-active-dormant-or-what/.

Thao. Digital Museum of Taiwan Indigenous People. Retrieved from http://www.dmtip.gov.tw/web/en/page/detail?nid=13.

The Baruklinting Dragon. (2016.) Indonesian Folklore. Retrieved from http://indonesianfolklore.blogspot.com]/2007/10/ baruklinting-dragon-folklore-from.html.

The Bones of the Wawel Dragon. Atlas Obscure. Retrieved from www.atlasobscure.com/places/the-bones-of-the-wawel-dragon.

The Candle Dragon. (2014.) Cultural China. Retrieved from http://traditions.cultural-china.com/en/13Traditions1298.html.

The Candle Dragon. Myths and Legends. Retrieved from creator.myths.e2bn.org/show/4722.

The Candle Dragon. (2017.) USC Digital Folklore Archives: A database of folklore performances. Retrieved from http:// folklore.usc.edu/?p=36521.

The Cuelebre. Draconika. Retrieved from http://www.draconika.com/legends/cuelebre.php.

The Cuelebre. (2015, August 31.) Valley of Dragons. Retrieved from http://valleyofdragons.com/the-cuelebre/.

The curse of Nyaminyami. (2003, December) Faces: People, Places, and Cultures. Retrieved from http://link. galegroup.com/apps/doc/A112538912/ITOF?u=mlin_w_umassamh&sid=ITOF&xid=f69395ba.

The Dreaming. (2015, March 31.) Australian Government. Retrieved from http://www.australia.gov.au/ about-australia/ australian-story/dreaming.

The Jawoyn People. Nitmiluk Tours: Jawoyn 'Sharing Our Country.' Retrieved from https://www. nitmiluktours.com.au/ about-us/jawoyn-people.

The legend of Drachenfels (Dragon Rock). (2015, November 19.) Middle Europe. Retrieved from www. middle-europe.cz/the-legend-of-drachenfels-dragon-rock.

The Legend of Nyami Nyami. (2014, March 21.) Victoria Falls 24.com. Retrieved from http://victoriafalls24. com/blog/2014/03/21/the-legend-of-nyami-nyami/.

The Legend of Sun-Moon Lake. (1996, September 7.) Folk Stories of Taiwan. Retrieved from http://www. taiwandc.org/folk-sun.htm.

The Legend of the Drachenfels; Or, How the Fell Dragon Fell. (1874). Magenta 4(1), 9-9.

The Legendary Origins of the Viet People. Vietnam Culture: Brings Vietnamese Culture to the Rest of the World. Retrieved from http://www.vietnam-culture.com/articles-47-4/The-Legendary-origins-of-the-Viet-people.aspx/.

The Moon Temple or Amaru Machay. Ancient Mysteries Explained. Retrieved from http://www.ancient-mysteries- explained.com/moontemple.html.

The Origin of Viet People. (2013, April 21.) Vietnamese Myths and Legends. Retrieved from https://sites. google.com/site/ vietnamesemythsandlegends/the-origin-of-viet-people.

The Story of Bathala. (2010, November 29.) Bathala. Retrieved from http://bathalanglangitatlupa.blogspot. com/2010/11/ story-of-bathala.html.

The Story of Bathala (Luzon Creation Myth). Tingin-tingin din! Retrieved from http://tingintingindin.weebly. com/ uploads/1/8/3/1/18312609/pre-spanish_creation_stories.pdf.

The Three Treacle Wells of Longwitton. (2011, March 18.) Leaves in the Lonnen. Retrieved from https:// fettlereetly.wordpress.com/2011/03/18/the-three-treacle-wells-of-longwitton/.

Tofighian, Nadi. (2008.) José Nepomuceno and the creation of a Filipino national consciousness. Film History 20(1), 77-94.

Transactions of the Asiatic Society of Japan, Volumes 17-18. The Gospel in All Lands.

Trip Mo, Trip Ko. (2012.) University of California Los Angeles Light and Dark.

Turner, Patricia and Charles Russell Coulter. (2000.) Dictionary of Ancient Deities. New York, NY: Oxford University Press.

Uktena-like monster incised on a Moundville pot. Retrieved from https://library-artstor-org.silk.library. umass.edu/asset/ARTSTOR_103_41822001453990.

Uwabami. (2019.) Yokai.com. Retrieved from http://yokai.com/uwabami/.

Valiente, Tito Genova. (2015 January 1.) A serpent, this earth and the end of the year. Business Mirror. Retrieved from http://search.proquest.com/docview/1644507809.

Van Ky, Nguyên. (2002.) Rethinking the Status of Vietnamese Women in Folklore and Oral History. In ViêtNam exposé : French scholarship on twentieth-century Vietnamese Society. Gisele Bousquet and PierreBrocheux, Eds. Ann Arbor, Michigan: The University of Michigan Press.

Vandale, James. (1999, March 11.) The Drac: French Tales of Dragons and Demons. Reptilian Agenda. Retrieved from http://www.reptilianagenda.com/myth/m110399b.shtml.

Varner, Gary R. (2009.) Sacred Wells: A Study in the History, Meaning and Mythology of Holy Wells. Algora Publishing.

Viasova, Eugenia. (2011, October 21.) Russian Dragon. Proper Russian. Retrieved from blog. properrussian.com/2011/10/russian-dragon.html.

Vogel, Jean Philippe. (1926.) Indian Serpent-lore: Or, The Nagas in Hindu Legend and Art. New Delhi, India: Asian Educational Services.

Von Grunebaum, G.E. and Roger Caillois. (1966.) The Dream and Human Societies. Berkeley, CA: University of California Press.

Wallace, Howard. (1948.) Leviathan and the Beast in Revelation. The Biblical Archaeologist 11(3), 61-68.

Walsh, T. F. (2013, February 11.) Mythology Monday: Zmey, The Slavic Dragon. Retrieved from https:// tfwalsh.wordpress.com/2013/02/11/mythology-monday-zmey-the-slavic-dragon/.

Walton, Evangeline. (2004.) The Mabinogion Tetralogy. New York: The Overlook Press.

Wang, Andrea. (2016.) The Nian Monster. Kirkus Media LLC.

Warner, Elizabeth. (2002.) Russian Myths. Austin, TX: University of Texas Press.

Wawel. Wawel.krakow. Retrieved from www.wawel.krakow.pl/en/index.php?op-11.

Werner, E. T. C. (1922, 1994.) Myths and Legends of China. Dover Publications.

What Is the Legend of Nian? (2015.) Quora. Retrieved from https://www.quora.com/What-is-the-legend-of-Nian.

Whittall, Austin. (2012.) Monsters of Patagonia. Ushuaia: Zagier & Urruty Publications.

Wianecki, Shannon. The Sacred Spine. Maui Magazine. Retrieved from http://mauimagazine.net/the-sacred-spine/.

Wikipedia and Livres Groupe. (2011.) Reptile Legendaire : Dragon Legendaire, Naga, Serpent Legendaire, Uraeus, Lindworm, Dragon Oriental, Tarasque, Dragon Occidental, Basilic. Books LLC, Wiki Series.

Wilbert, Johannes and Simoneau, Karin, Eds. (1984.) Folk Literature of the Tehuelche Indians. Los Angeles, California: UCLA Latin American Center Publications.

Wilkinson, Carole. The Foolish Dragon. The Dragon Companion. Retrieved from www.carolewilkinson.com. an/ dragoncompanion/thefoolishdragon.php.

Williams, George M. (2003). Handbook of Hindu Mythology. Santa Barbara, California: ABC-CLIO, Inc.

Willoughby, Charles C. (1935.) Michabo the Great Hare: A Patron of the Hopewell Mound Settlement. American Anthropologist 37(2), 280-286.

Wolfe, David Michael Wolfe. Legend of the Tlanuhwa and the Uhktena. Cherokees of California. Retrieved from http:// www.powersource.com/cocinc/articles/tlanuhwa.htm.

Wolfgramm, Emil. (1993.) Comments on a Traditional Tongan Story Poem. Manoa 5(1), 171-175.

Worms, E. A. (1955.) Contemporary and Prehistoric Rock Paintings in Central and Northern North Kimberley. Anthropos 50(4/6), 546-566.

Ye, Chao, Chen, Ruishan, and Chen, Mingxing. (2016.) The impacts of Chinese Nian culture on air pollution. Journal of Cleaner Production 112, 1740-1745.

Yolen, Jane. (2014.) Favorite Folktales from Around the World. Knopf Doubleday Publishing Group.

Yu, Insun. (1999.) Bilateral social pattern and the status of women in traditional Vietnam. South East Asia Research 7(2), 215-231.

Zhan, Jade. (2019.) Living By the Numbers. Shen Yun Performing Arts. Retrieved from https://www.shenyun.com/blog/ view/article/e/GVxkb2N-pEY/chinese-lucky-numbers.html.

Zhao, Qiguang. (1989.) Chinese Mythology in the Context of Hydraulic Society. Asian Folklore Studies 48(2), 231-246.

Zhishu, F. (2017.) A Chinese renaissance. Nature Plants 3(1).

Zmaj and the Dragon Lore of Slavic Mythology. (2015, January 5.) Ancient Origins: Reconstructing the Story of Humanity's Past. Retrieved from www.ancient-origins.net/myths-legends-europe/zmaj-and-dragon-lore-slavic-

mythology-002984.

Zmey Gorynych. (2011, December 5.) Dragon's Corner: Exploring dragonkind around the world! Retrieved from https://dragonscorner.wordpress.com/2011/12/05/zmey-gorynych.

Zoroastrianism. (2016, April 4). New World Encyclopedia, Retrieved from http:// www.newworldencyclopedia.org/p/index.php?title=Zoroastrianism&oldid=995084.

Index

F

G

H

I

J

K

L

M

N

W

X

Y

Z

Made in United States
North Haven, CT
19 December 2023

46170823R00142